How To Pass Your FAA Part 107 Pilot Exam

RAOUL WATSON

How To Pass Your FAA Part 107 Pilot Exam

Raoul Watson

ISBN: 979-8-218-97791-7

TABLE OF CONTENTS

PREFACE..1

    Who am I?   2

    Why did I decide to write this book?   2

CHAPTER I 5

    Materials Needed   5

    One Site Repository and access code   6

    Regulatory   7

    Exam Description & Practice Regulations   16

    Regulations Practice Exam   20

CHAPTER II 27

    Crash Course to The Aviation World   27

    The Coordinate System   35

    Altitude & Speed   42

    Reading Charts   43

CHAPTER III 49

    Controlled Airspace   49

    Class B Airspace   50

    Class C Airspace   51

    Class D Airspace   52

    Class E Airspace   52

    Uncontrolled Airspace   53

    Class G Airspace   53

    Special Use Airspace   53

    Restricted, Prohibited, Warning, and MOAs   54

    Temporary Flight Restrictions (TFRs) and NOTAMs   57

CHAPTER IV...63

Aviation Weather   63

Atmospheric Pressure   64

Dew Point and Air Temperature   64

Relative Humidity   64

Density Altitude   65

Fog and Classifications   65

Ceiling & Visibility   67

Precipitation & Thunderstorms   68

Automated Weather Reports   69

METAR and TAF Abbreviations & Acronyms   75

CHAPTER V 79

Operations and Performance   79

Preflight Checklist   79

Load Factor   80

Operation Above Human Beings   83

Emergency Procedures   85

Communications   85

Aeronautical Decision Making (ADM)   88

Best Practices   92

CHAPTER VI..97

Getting Ready for The Test   97

Registering with the FAA   97

The Day of The Test   101

Submitting Test Results and Applying for The Certificate   104

AFTERWORD 107

PREFACE

I was thrilled when the Federal Aviation Administration (FAA from here on) decided to integrate drones into the National Air Space system. It's about time they are recognized as players in our skies. This was in August of 2016. I became discouraged when I learned the requirements of becoming an FAA-certified Remote Pilot. Many people believed that the study and test requirements were just overkill. The knowledge required doesn't seem to match the needs of the industry. After all, when do drone pilots need to look at aeronautical maps depicting navigation aids, runways, and radio communications? Unfortunately, if you want to be on the field playing with other airmen, we must demonstrate that we know the rules and regulations to keep all the players safe. In addition, our drones are actually named "Unmanned *Aircraft* System" or UAS. Notice the word "aircraft." Since it is an aircraft, it is regulated like any other aircraft by the FAA, which is the body of authority for aviation in the United States. My goal with this book is to consolidate the knowledge most relevant to drone operation and provide all the material needed to pass the exam. Even if you have zero aviation knowledge, by the end of this book, you will gain a brand-new skill set that you can be proud of.

WHO AM I?

I am an FAA-certified Part 107 Remote Pilot who, at one time, was in your position. I understand the struggle and confusion navigating through the plethora of information preparing for the exam. I also understand the challenges with the study materials, which can be foreign for those new to aviation. I was one of the first batch of Remote Pilots in the United States. The Part 107 Certificate was established in August of 2016. At the time I applied, there were no official remote pilots. The requirement then was a lot more vigorous than today. There were FAA courses you had to take before you could register for the exam, and then two years later, you had to come back and show the FAA that you were still current with your knowledge. I had to take my exam at La Guardia airport with the other pilots at the FAA exam center. They had never even heard of part 107 at that time. Thank goodness many of those requirements have been made less stringent and more manageable for us.

WHY DID I DECIDE TO WRITE THIS BOOK?

I met many drone lovers who would like to take their hobby to the next level by providing services to many industries that need them. Among others, real estate, inspection, and survey services. In addition, many could not have flown their drones in their location without being a Part 107 pilot. When I was in Rockaway Beach, New York, I could never fly my drone without this certificate because my location was under a very strict controlled airspace. Rockaway Beach is in a direct path of aircraft approaching for landing at JFK airport runway 4 left and 4 right. The people I met are primarily frustrated with the material they need to gather to study. Wouldn't it be nice to have them all in one place, organized logically according to what will be required to take the Part 107 test?

Undoubtedly, there is so much material to cover that this is not for the faint-hearted. I have tried not to condense this book and omit the information you need for the FAA exam.

Like any other knowledge-based work, I did not develop this book in a vacuum or accumulate my knowledge alone. When it comes to flying, every day is a learning experience for me. I want to thank all the remote pilots and organizations who provided the knowledge found in this book. I want to thank the lawmakers who made laws to integrate drones into the National Airspace System (NAS), enabling us to fly our drones and improving their operations' safety.

Finally, thank you for your willingness to use this book, a publication that might help you accomplish your goal of becoming a Part 107 Remote Pilot. I promise I will not waste a single minute of your time. If it is in this book, you can be assured you need the information and that it will come up on the FAA exam.

Edition Check: The field of drones and its regulations and laws is very dynamic. If you find this book many years after its publication, you should see if a newer edition is available.

Please don't ignore the Assignments at the end of each chapter. The subjects we study are vast, and the assignments give you additional resources to gain knowledge.

This page intentionally left blank

CHAPTER I

MATERIALS NEEDED

Before we get started, we must get ourselves organized and gather the materials needed to utilize our time effectively. You will be accessing tons of information from different websites, and you will be required to create accounts with several FAA departments, which you will need to register your drone, take the exam, and apply for a certificate. A plethora of regulatory or informational information must be accessed on websites. And while it would be convenient to include them here, the volume would be prohibitive. In addition, we should not rely on material locally when it is dynamic. Therefore, we will access them online to ensure we get the most up-to-date information. So, get yourself a notepad or create a file to jot down website locations, user IDs you will be creating, and passwords. You must keep track of identification numbers issued to you, which will remain with you for the rest of your remote pilot endeavor. I also strongly suggest that you study in front of a computer since our phones, albeit convenient, don't have the visual real estate needed to put several resources side by side to read and study.

ONE SITE REPOSITORY

When I started this book, my main concern was making your time spent as efficiently as possible. The biggest challenge you have is the need to write down so many website addresses and information sources that I decided to create a simple one-page site where all the links are listed, and all you must remember is <u>one site</u>. You can try it now by launching your browser. Type this in your browser navigation box "raoulwatson.com/107". You can also scan the QR code with your phone camera and click the link shown on the camera. Even though this is available, I will still list the complete website URLs in the book text. The OneSite not only has the complete resources mentioned in this book but also has an area for you to practice for the exam. There are two choices: use the FAA site or use the test I have designed. The FAA site gives you 46 questions, always the same, not randomized in occurrence. My exam covers each chapter, and it's randomized from a collection of questions database I have compiled. The exam does require an access code of "book107" (all lowercase) to gain access.

How To Pass Your Part 107 Pilot Exam
Links referred in the book:

Study Material and References:

The law governing part 107, (Code of Federal Regulation or CFR) Title 14, Chapter I, subchapter F, Part 107

Part 89 Remote ID Regulation (CFR)

Subchapter A, Part 91
(General Operating and Flight Rules)

Subchapter C, Part 47
(Aircraft Registration CFR)
47.14 Serial Number for UAS
48   Marking, Registration of UAS

Personal Resources & Study Material:

Aeronautical Charts and Airspace

Get METAR

METAR/TAF decoder

Get TFR-NOTAM

Briefing at 1800wxbrief.com

FAA Safety Team (Courses and Webinars)

FAA Aeronautical Chart Users' Guide

FAA Pilot's Handbook of Aeronautical Knowledge

FAA Aeronautical Information Manual (AIM)

Testing Material:

Testing supplement book referred to in FAA exam

FAA Practice Exam
(select "Exam Resources, scroll down and find "Unmanned Aircraft General - Small (UAG) and click "Sample Test)

Book Practice Exams
(enter your name and access code from the book)
Knowledge Quiz (Regulations)
Knowledge Quiz (Airspace)
Knowledge Quiz (Weather)
Knowledge Quiz (Operations)
** PART 107 – All subjects **

If you don't have it, you can download the latest Adobe Acrobat PDF reader for free.
Download Adobe Reader Installer

Drone specific

Drone Registration

UAS Facility Map

LAANC and providers

Remote Pilot Test Application

Test Registration (IACRA)

Schedule to take a test (PSI Test)

Contact the author watsonr@intelligencia.com

REGULATORY

So why is it called part 107, some may wonder? In 2016, when it all began, a sleuth of rules and regulations were put into law known as the Code of Federal Regulation (CFR); specifically, the law is placed into Title 14 (Aeronautics and Space) in Chapter I, subchapter F, Part 107. So, this is where we get "part 107" from. The rules and regulations can be found in this CFR, located at

https://www.ecfr.gov/current/title-14/chapter-I/subchapter-F/part-107.

Remember to use the OneSite page so you don't have to type the link above. Have the regulations open and available as you review this chapter.

In addition, there are other CFRs applicable to drone operations, namely.

Part 47:

(https://www.faraim.org/faa/far/cfr/title-14/part-47/index.html#seqnum47.14)

Part 48:

(https://www.ecfr.gov/current/title-14/chapter-I/subchapter-C/part-48)

part 89:

(https://www.ecfr.gov/current/title-14/chapter-I/subchapter-F/part-89)

You should familiarize yourself with the CFRs by simply doing a light reading to give you an idea of what it is about. All test questions answer basically just a quotation of the law (we see more of it later). From now on, whenever you see brackets, e.g. [107.37], that would be the reference to the CFR. We begin with eligibility to apply for Part 107 Remote Pilot Certification, and we will conclude with sample exam questions. Note: Items in *italics* are the author's comments:

[107.61] Eligibility

At least 16 years old.
Be able to read, speak, write, and understand English.
Be in a physical and mental condition to safely fly a drone.
Pass the initial aeronautical knowledge exam:
Unmanned Aircraft General – Small (UAG).

The aeronautical knowledge exam, "Unmanned Aircraft General—Small (UAG)," consists of 60 questions and requires 70% to pass.

Now we will continue with the regulations specific to drone operating limitations. At the end of the chapter, take the knowledge exam for regulations at the OneSite.

Acronyms/Abbreviations used:

PIC   Pilot In Command (who's at the remote controller)

VO   Visual Observer (one assisting the PIC)

sUAS   small Unmanned Aircraft System (the drone)

MPH   Miles Per Hour (speed)

KT   Knots, speed used in aviation (approx. 1.15 mph)

AGL   Above Ground Level

FRIA   FAA Recognized Identification Areas

CBO   Community-Based Organizations

RID   Remote Identification ID for drones

[107.51]

The ground speed of sUAS may not exceed 87 kt (100 mph).

The altitude of the sUAS may not exceed 400 feet unless the sUAS is within 400 feet of a structure in which case the sUAS may fly up to 400 feet above the structure. *While this is in the regulation, in controlled airspace, your clearance from ATC may disallow this.*

The operating visibility, as observed from the PIC, is at least 3 statute miles.

The minimum distance of the sUAS to the clouds must not be less than 500 feet below, and 2000 feet horizontally.

[48.15]

Unless the sUAS weighs 0.55 pounds or less on take-off (including everything on board and attached) the sUAS must be registered and marked/labeled.

[48.25]

The sUAS must be registered by its owner using the legal name of the owner unless the owner is less than 13 years old, then the sUAS must be registered by a person who is at least 13 years of age.

[48.200] and [48.205]

The sUAS must be marked or labeled legibly on the external surface of the sUAS with the registration number obtained from the FAA drone registration. *It's not a bad idea to also include your phone number in case the drone is lost and found by someone.*

[107.31] and [107.33]

The sUAS must be within the visual line of sight (unaided by any device except for corrective lenses) of the PIC or the VO, if one is used. If a VO is used, the VO must also be able to communicate effectively with the PIC.

[107.29]

No person may operate an sUAS at night unless the PIC has a valid Remote Pilot certificate or recurrency after April 6, 2021, and the sUAS is equipped with anti-collision lighting visible for at least 3 statute miles. *(Exam and recurrency after April 2021 include night flight regulations questions).*

[107.29]

No person may operate a small unmanned aircraft system during civil twilight unless the small unmanned aircraft has lighted anti-collision lighting visible for at least 3 statute miles that has a flash rate sufficient to avoid a collision. The remote pilot in command may reduce the intensity of, but may not extinguish, the anti-collision lighting if he or she determines that, because of operating conditions, it would be in the interest of safety to do so.

Civil twilight refers to the following:

A period that begins 30 minutes before official sunrise and ends at official sunrise; and

A period that begins at official sunset and ends 30 minutes after official sunset.

In Alaska, the period of civil twilight is defined in the Air Almanac.

[107.25]

No person may operate a small unmanned aircraft system from a moving land or water-borne vehicle unless the small unmanned aircraft is flown over a sparsely populated area and is not transporting another person's property for compensation or hire.

[89.110]

All registered sUAS may not be flown after September 16, 2023, unless equipped to broadcast the remote identification ID (RID). The remote ID must be broadcast from takeoff to landing. If the UAS is no longer broadcasting the remote ID, the PIC must land it as soon as possible.

There is an exemption to this rule, namely an area called FRIA. The FAA usually approves this area upon request from COBs such as RC Flying clubs, Educational Institutions, etc. In this specific area, drones can be flown without RID. The location of FRIAs can be found using the link on the OneSite website under the UAS facility map.

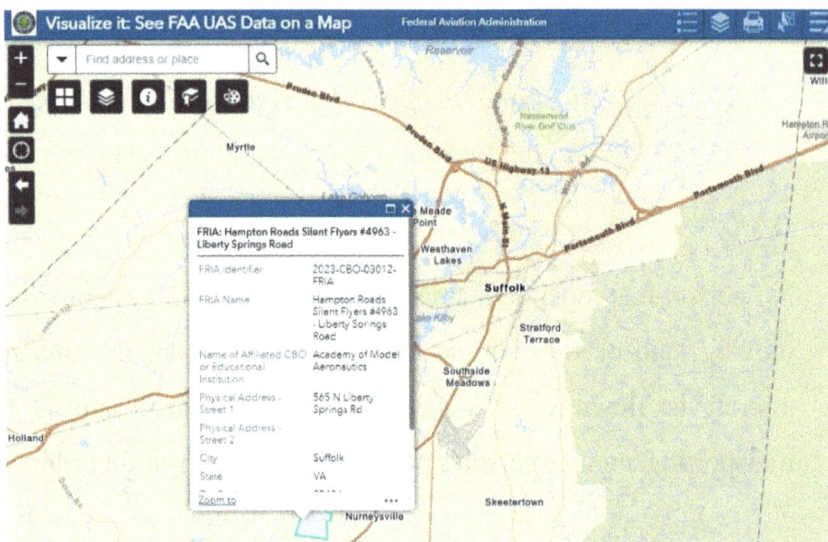

A FRIA area is identified by clicking a location on the map

There is another CFR mentioned in Part 107 [107.27]. While this CFR is not drone-specific, since our drones are classified as *aircraft*, it applies to us as well. If you are a PIC or VO, you must comply with this provision.

This CFR is in Subpart A of Part 91.

https://www.ecfr.gov/current/title-14/chapter-I/subchapter-F/part-91/subpart-A

[91.15]

No PIC may allow any object to be dropped from the aircraft in flight that creates a hazard to persons or property. However, this section does not prohibit the dropping of any object if reasonable precautions are taken to avoid injury or damage to persons or property.

[91.17]

No PIC or VO may conduct drone operation under the influence of alcohol or within 8 hours after consumption or when their blood alcohol level is 0.04 or greater. *This is the regulation for aircraft pilots known as "eight hours from the bottle to the throttle."*

[91.19]

You may not attach any drugs or narcotics to a UAS unless it is authorized by any Federal or State agency. *This is for possibly automated prescription drug delivery.*

[107.57]

Violations or convictions related to [91.17] or [91.19], including refusal to submit to an alcohol test, may be used as grounds for suspension or revocation of your Remote Pilot certificate and can be used as a base to deny your application.

[107.41] and [107.45]

No person may operate a small unmanned aircraft in Class B, Class C, or Class D airspace or within the lateral boundaries of the surface area of Class E airspace designated for an airport unless that person has prior authorization from Air Traffic Control (ATC). You also may not operate in prohibited or restricted areas unless you have permission from the using or controlling agency.

[107.37] and [107.43]

Each small unmanned aircraft must yield the right of way to all aircraft, airborne vehicles, and launch and reentry vehicles. Yielding the right of way means that the small unmanned aircraft must give way to the aircraft or vehicle and may not pass over, under, or ahead of it unless well clear. No person may operate a small unmanned aircraft so close to another aircraft as to create a collision hazard. You are also prohibited from causing interference with the operation at any airport, heliport, or seaplane base.

There are many other regulations, but they will be covered in the individual-related sections. For example, we will discuss Regulations related to the operation of sUAS in the "Operations" section and regulations about Airspace in the "Airspace" section.

Numbers Review

Studies suggest that many individuals have greater problems remembering numbers inside sentences.[1] This is a quick snapshot of numbers related to the regulation area.

16   eligibility bottom age for the Remote Pilot certificate.

10   within days to report accidents when required.

87   the max speed limit in knots for the drone.

400   max altitude for drones in feet.

500   below cloud base vertical separation distance and the base amount of damage to file an accident report.

2000   horizontal separation distance to the cloud.

0.55   weight bottom limit for registration.

13   age bottom limit for registration.

30   minutes before sunrise and after sunset (civil twilight).

8   hours after consumption of alcohol.

0.04   base blood alcohol level.

11   impact force for category 2.

25   impact force for category 3.

55   upper limit weight of sUAS

---

[1] Psychophysical Study of Numbers, John C. Baird & Elliot Noma, Dartmouth College, Hanover, New Hampshire. Psychol. Res. 37, 281-297 (1975)

EXAM DESCRIPTION

The Part 107 exam is broken down into several categories:

1. <u>Applicable regulations</u> relating to sUAS rating privileges, limitations, and flight operation.

2. <u>Airspace</u> classification and operating requirements, and flight restrictions affecting sUAS aircraft operation.

3. <u>Weather</u> and the effects of weather on sUAS performance.

4. <u>Operations and Performance:</u>

   - Small unmanned aircraft loading and performance.

   - Emergency procedures.

   - Radio communication procedures.

   - Physiological effects of drugs and alcohol.

   - Aeronautical decision-making and judgment.

   - Crew resource management.

   - Airport operations.

   - Maintenance and preflight inspection procedures.

   - Operation at night.

This book is organized based on the categories applied in the exam. At the end, I will walk you through the entire process of applying for the exam, what to do during the exam, and, upon passing, a follow-up request for the Remote Pilot certificate.

From past experiences, the FAA breaks down the exam into five major categories with the following approximation:

Regulations (20%)   +- 12 questions

Airspace Classification (20%)   +- 12 questions

Weather (15%)   +- 9 questions

Loading & Performance (5%)   +- 3 questions

Operations (40%)   +- 24 questions

During the exam, you will be given a testing supplement book with material referred to in the exam. For example, a test question may ask you to refer to Figure 24 in the book. You will need to look at this figure in order to answer the question. We will get you used to referring to this book in our discussion. The link for the test supplemental book is available on the OneSite:

We will often refer to this book to answer test questions just as you would during the real test.

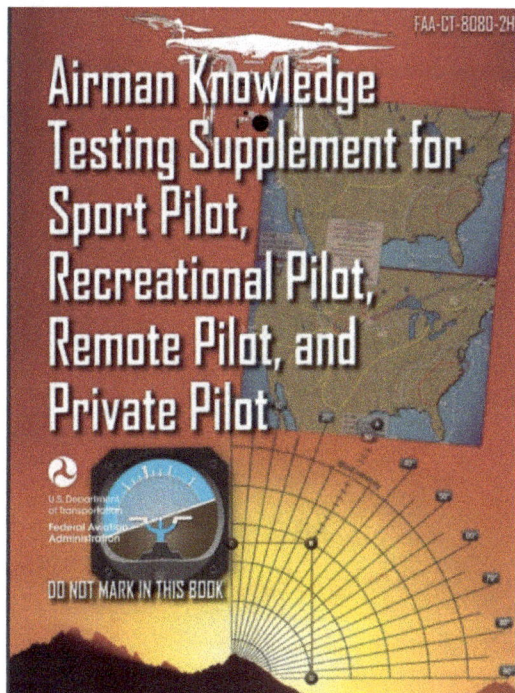

This book is in PDF format (Adobe Acrobat) file, so you need Adobe reader (available to download for free from the OneSite webpage. Here is the official link for this FAA test supplement book:

https://www.faa.gov/sites/faa.gov/files/training_testing/testing/supplements/sport_rec_private_akts.pdf

If you use the OneSite, the book will open in your browser, which likely would be able to display PDF files properly. Some of the pages have images in landscape, or printed sideways as you open the book. During the real exam, there is no problem since you could just rotate the book. In a browser or Adobe Reader, however, we must rotate the page using the rotate icon. Here are samples of the rotate icon. Keep in mind that if only one directional rotation is available, you may have to rotate it several times until you get the correct orientation for reading.

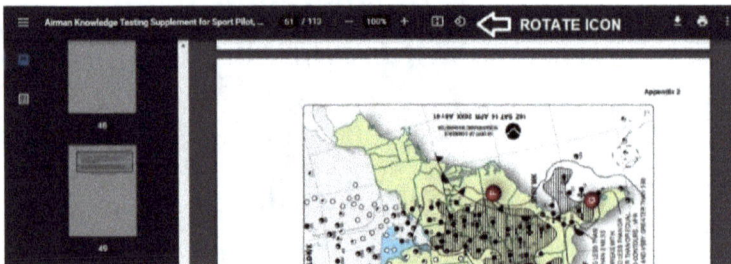

Adobe
Reader

We will practice the use of this book. Both during the exam and practice, you must use this book efficiently without a waste of time. During the exam, always utilize the table of contents (usually the first 7th and 8th page, insert a piece of paper here). When the question calls for Figure 24 for example, immediately locate this figure from the Table of Contents. This will save you time flipping pages during the exam. On the computer, our life is a little easier since you can press Ctrl-F simultaneously (for find), type "figure 24" in the search field, and hit Enter.

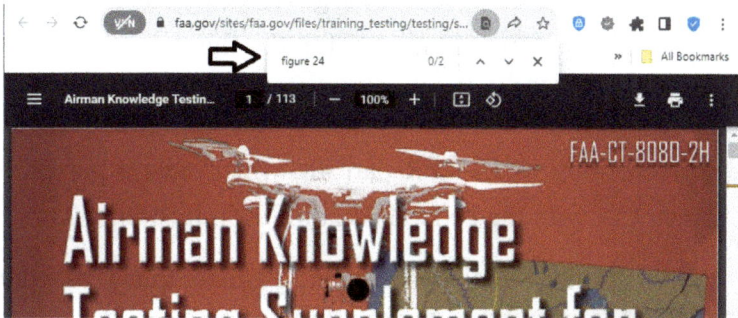

You will most likely have to press Enter more than once since the first one will find the text in the Table of Contents. Simply press Enter again.

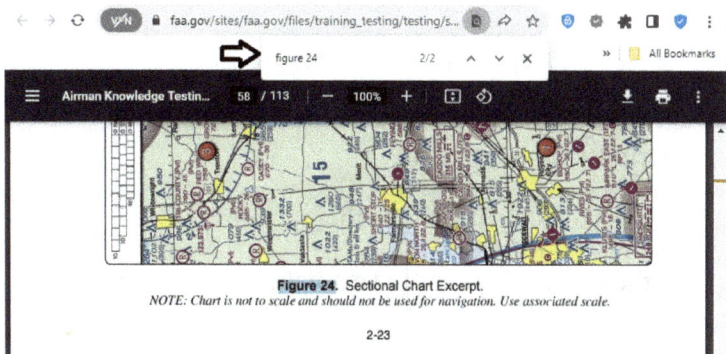

Figure 24. Sectional Chart Excerpt.
NOTE: Chart is not to scale and should not be used for navigation. Use associated scale.

2-23

REGULATIONS PRACTICE EXAM

1. Under what condition would a small UA not have to be registered before it is operated in the United States?

A. When the aircraft weighs less than 0.55 pounds on takeoff, including everything that is on board or attached to the aircraft.

B. When the aircraft has a takeoff weight that is more than 0.55 pounds, but less than 55 pounds, not including fuel and necessary attachments.

C. All small UAS need to be registered regardless of the weight of the aircraft before, during, or after the flight.

----

2. According to 14 CFR part 48, when must a person register a small UA with the Federal Aviation Administration?

A. All civilian small UAs weighing greater than 0.55 pounds must be registered.

B. When the small UA is used for any purpose other than as a model aircraft.

C. Only when the operator will be paid for commercial services.

----

3. According to 14 CFR part 48, when would a small UA owner not be permitted to register it?

A. All persons must register their small UA.

B. If the owner does not have a valid US driver's license.

C. If the owner is less than 13 years of age.

4. According to 14 CFR part 107, how may a remote pilot operate an unmanned aircraft in Class C airspace?

A. The remote pilot must monitor the Air Traffic Control (ATC) frequency from launch to recovery.

B. The remote pilot must have prior authorization from the Air Traffic Control (ATC) facility having jurisdiction over that airspace.

C. The remote pilot must contact the Air Traffic Control (ATC) facility after launching the unmanned aircraft.

----

5. According to 14 CFR part 107, what is required to operate a small UA within 30 minutes after official sunset?

A. Use of anti-collision lights.

B. Must be operated in a rural area.

C. Use of a transponder.

----

6. A small UA causes an accident, and your crew member loses consciousness. When do you report the accident?

A. No accidents need to be reported.

B. When requested by the UA owner.

C. Within 10 days of the accident.

7. According to 14 CFR part 107, what is the maximum ground speed for a small UA?

A. 87 mph.

B. 87 knots.

C. 100 knots.

----

8. Upon request by the FAA, the remote pilot-in-command must provide

A. a logbook documenting small UA landing currency.

B. a remote pilot certificate with a small UAS rating.

C. any employer-issued photo identification.

----

9. When may a remote pilot reduce the intensity of an aircraft's lights during a night flight?

A. At no time may the lights of an sUAS be reduced in intensity at night.

B. When a manned aircraft is in the vicinity of the sUAS.

C. When it is in the interest of safety to dim the aircraft's lights.

10. The refusal of a remote PIC to submit to a blood alcohol test when requested by a law enforcement officer

A. is grounds for suspension or revocation of their remote pilot certificate.

B. can be delayed for a period of up to 8 hours after the request.

C. has no consequences to the remote pilot certificate.

----

11. When may a person who does not hold a remote pilot certificate operate a UAS for hire?

A. When directly supervised by a visual observer.

B. Only when the flight operations have been approved by a certified UAS pilot.

C. When under direct supervision of a Remote PIC who is immediately available to take control if necessary.

----

12. Provided no property is carried for compensation or hire, an operator may fly a UAS from a moving vehicle if the UAS is:

A. Equipped with a transponder.

B. Operated over a sparsely populated area.

C. Equipped with visible anti-collision lights.

----

13. How soon must a Remote PIC report an accident reportable to the FAA?

A. Within 30 days.

B. Within 10 days.

C. Within 14 days.

14. While surveying a building, your UAS lost control and crashed hitting a person. The person requires hospitalization, but the injury can fully heal (including, but not limited to, head trauma, broken bone(s), or laceration(s) to the skin that requires suturing).

A. This accident needs to be reported within 10 days.

B. This accident does not have to be reported.

C. This accident needs to be reported within 14 days.

----

15. While surveying a building, your UAS lost control and crashed, hitting a parked car damaging the side mirror costing $400. The local repair shop charges $600 to repair the mirror.

A. This accident must be reported because it costs over $500.

B. This accident does not need to be reported because the value is less than $500.

C. This accident does not need reporting because it's covered by car insurance.

Assignments:

1. **Review** CFR Title 14, Part 107 (use the OneSite).

2. Take the Knowledge Quiz "Regulations" at the OneSite. Remember the access code is "book107".

ANSWER KEY TO REGULATIONS PRACTICE EXAM

Chapter 1

1. A   [48.15]
2. A   [48.15]
3. C   [48.25]
4. B   [107.41]
5. A   [107.29]
6. C   [107.9]
7. B   [107.51]
8. B   [107.7]
9. C   [107.29]
10. A   [107.59]
11. C   [107.12]
12. B   [107.25]
13. B   [107.9]
14. A   [107.9]
15. B   [107.9]

This page intentionally left blank

CHAPTER II

A CRASH COURSE TO THE AVIATION WORLD

Looking at what was included in the exam above, some of you may think, "What have I gotten myself into?" I can assure you that you should not be discouraged. You can do this. Even if you have no idea about aviation terminology and knowledge, by the time you finish this book, you will have a new skill set and be well prepared. All it takes is determination and open-mindedness to learn something new.

We will divide this crash course chapter into sections specific to the knowledge and skills required for the exam. The major sections in this crash course are:

- Compass Direction.

- Coordinate System.

- Altitude.

- Speed.

- Reading Aeronautical Charts.

- Aviation terminology and chart symbols.

Compass Direction

The compass direction is used to show your relative direction to determine direction relative to the earth's magnetic poles. We all know the cardinal directions of North, West, South, and East. The concept behind it is based on the fact that the earth's circumference is a circle. A circle consists of 360 degrees. So, in a compass, the circle is divided by 360. Here is how a compass rose looks:

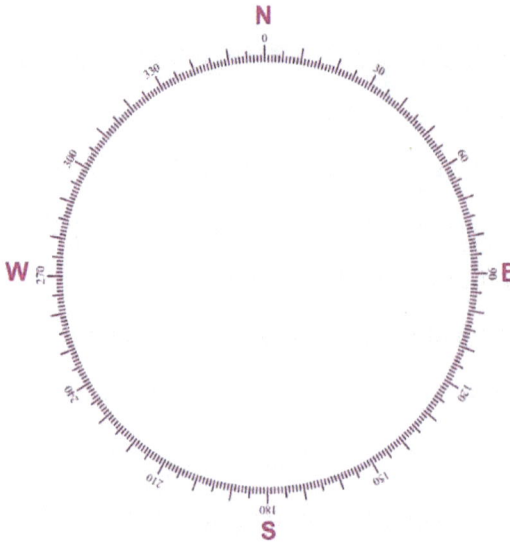

Basically, a circle divided by 360 numeric labels, beginning with zero and going clockwise. The direction where an object moves on Earth is called "heading." So, if my heading is 90°, I am moving East. 180°, I am moving South, and so on. The device that is used to display the direction is the compass. This is one instrument that is critical for navigation in the aviation world. In the aviation world, runways at airports are numbered based on their heading, rounded to the nearest ten, with the last digit dropped. For example, if a runway has a heading of 224 degrees, then this number would be rounded to 220. Dropping the last digit, this runway would be runway 22. This is the large number painted at the end of the runway; it's called "Runway Designator." So, the marking on the opposite end of the runway would be the heading in the opposite direction.

Opposite number math: If the runway number is greater than 18, subtract 18 to find the opposite marking. If it is 18 or less, then add 18. For example:

Runway 22 (greater than 18), opposite is $22 - 18 = 4$.

Runway 13 (18 or less), opposite is $13 + 18 = 31$.

Many large airports have parallel runways (runways next to each other). In such a case one would be referred to as "left" and the other "right" (also "center" if one exists). So, runway 22 would be either 22L or 22R. Same with the opposite of this runway, it will be 4L and 4R. Pop quiz: if one end of a runway is labeled 22L, what is the label on the opposite side? You must keep in mind that, on the other side, during the approach, this runway would be located on the right-hand side so that 22L would have 4R at the other end.

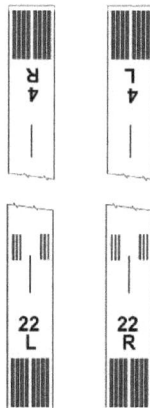

Understanding runways and orientation is an important skill since during the exam, you will find questions requiring this knowledge. For example, this one was on my FAA exam in 2016:

While monitoring the Cooperstown CTAF you hear an aircraft announce that they are midfield left downwind to RWY 13. Where would the aircraft be relative to the runway?

A) The aircraft is East.
B) The aircraft is South.
C) The aircraft is West.

Here, we encountered a couple of terms that we may not be familiar with: "CTAF" and "left downwind."

If you fly your drone in close proximity to an airport, you should monitor radio conversations so that you have a better situational awareness of your drone operation in relation to aircraft departing and arriving at this airport. If you do this a lot commercially, I recommend that you get a portable dual-channel receiver so that you can monitor the radio communication.

CTAF stands for "Common Traffic Advisory Frequency." It is commonly used at places such as uncontrolled small airports where ATC (Air Traffic Control) services are not provided. Pilots use CTAF to monitor and announce their intentions to give all the players better situational awareness. While prudent to monitor this, a drone pilot must never _transmit_ on this frequency.

In the United States, unless otherwise authorized, or indicated on a chart, the law requires "pilots of fixed-wing aircraft approaching to land must circle the airport to the left." (AIM 4-3-2 [b]). There are several components of a standard traffic pattern utilizing these terms:

1. Upwind leg. A flight path parallel to the landing runway in the direction of landing.

2. Crosswind leg. A flight path perpendicular to the landing runway beyond its takeoff end.

3. Downwind leg. A flight path parallel to the landing runway in the opposite direction of the landing.

4. Base leg. A flight path perpendicular to the landing runway beyond its approach end and extending from the downwind leg to the intersection of the extended runway centerline.

5. Final approach. A flight path in the direction of landing along the extended runway centerline from the base leg to the runway.

6. Departure. This is the flight path that begins after takeoff and continues straight ahead along the extended runway centerline. This climb continues at least half a mile beyond the runway.

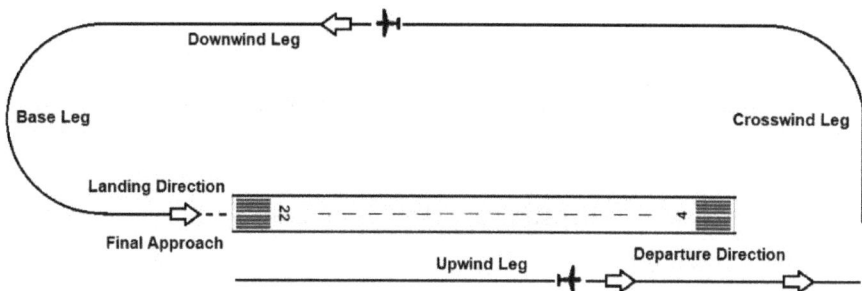

Aircraft usually depart and land into the wind because the airfoil lift characteristic is more efficient. It is also undesirable when the wind pushes you to a higher speed during landing.

So now we have this info, let's analyze the exam question:

> While monitoring the Cooperstown CTAF you hear an aircraft announce that they are midfield left downwind to RWY 13. Where would the aircraft be relative to the runway?
>
> A) The aircraft is East.
> B) The aircraft is South.
> C) The aircraft is West.

When you come to the test, they usually supply you with some blank note paper and a pencil. From our compass rose, we know the direction of the aircraft: a heading to runway 13 is around 130 degrees. Downwind indicates the aircraft is flying parallel to the runway in the opposite direction in preparation for a left-base approach. So, let's sketch the runway, pattern, and aircraft.

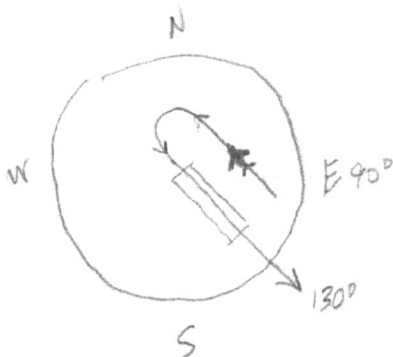

Hint: Use the paper provided during the test. If you have trouble visualizing the question, sketch it!

So, this question can be answered by the process of elimination since from our sketch, we can see clearly that the aircraft is neither South nor West of the runway. Any choice of Northeast or East fits the answer.

One subject adds a little confusion, especially during taking the FAA exam. This issue arises because, technically, there is more than one "north," namely true north, magnetic north, and geomagnetic north. For our study, we will concentrate only on what is relevant to the exam materials in the aviation world.

True North points to the geographical north pole, the northernmost point on the earth.

Magnetic North is the direction towards which any north-seeking device or magnet points. A compass in an aircraft, for example, seeks the magnetic north. The magnetic north shifts and changes over time in response to changes in the Earth's magnetic core.

Gosh, this is confusing! So how do we know whether the north is true or magnetic when it is mentioned? Here is the rule of thumb that I use:

If you can read it, and it is not about a compass, it's **true**. If you can hear it, it's **magnetic**. Remember this.

For example: You are reading a map or a weather report, true.

You hear a transmission "Cessna November five two niner two Romeo maintain heading one two zero." It's magnetic.

You saw a runway marking 22L from above. It's about compass direction, so it's not true but magnetic.

... I need to take notes and I don't have a piece of paper. OK...

_____

_____

_____

_____

_____

_____

_____

_____

_____

_____

_____

_____

_____

_____

_____

_____

_____

(You'll find some more in this book "Intentionally left blank" for you)

The Coordinate System

The coordinate system allows us to pinpoint any location using two sets of numbers: **latitude** and **longitude**. The earth's globe is divided by invisible lines going horizontally and vertically.

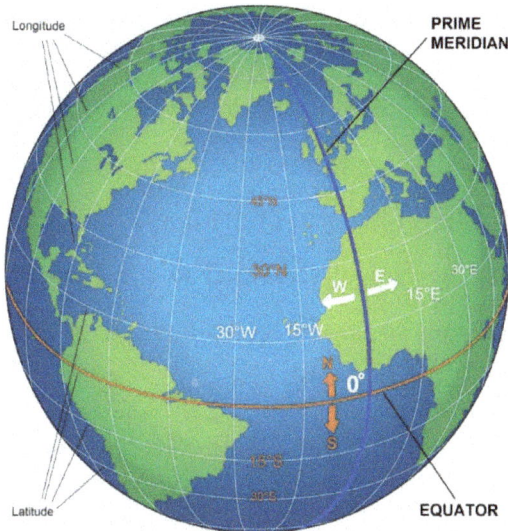

There are two very important lines, namely the prime meridian and the equator. The vertical blue line on the globe above is the prime meridian which runs through Greenwich, England. Or, to be more exact, 100 feet east of the Royal Observatory building on Blackheath Ave. This line invisibly divides the earth into the western and eastern hemispheres. **Longitude** is the name of the line at the prime meridian and all the others running vertically from the north pole to the south pole. It starts at 0 degrees at the prime meridian and increases one degree at a time (+1°) as you move eastward until you reach +180°, a line denoted as the international dateline where the date changes. It decreases one degree (-1°) as you move westward until -180° at the international date line. All longitudinal lines converge at the north pole in the north and the south pole in the south.

Since the earth's circumference is around 24,900 miles, the physical distance in one degree is roughly 69 miles (24,900 / 360°). Because the lines converge north and south, the distance of one degree on the longitudinal line varies from around 69 miles at the equator to near zero at the poles.

The red line in the previous globe picture is the equator, which invisibly divides the earth into the northern and southern hemispheres. **Latitude** is the name of the line at the equator and all the others running horizontally parallel to the equator. It starts at 0 degrees at the equator and increases one degree at a time (+1°) as you move northward until you reach +90° at the north pole. It decreases one degree (-1°) as you move southward until -90° at the south pole.

Unlike the longitudinal lines, the parallel lines would never converge, so the distance of one degree at the latitudinal line remains constant, roughly 69 miles.

Each degree on each line is divided further into minutes and seconds. Each degree consists of 60 minutes, denoted like the "feet" symbol ' (apostrophe), for example, 45'. Each minute has 60 seconds denoted with the "inches" symbol " (ditto mark), for example, 38". The distance for minutes is approximately 1.15 statute miles, and for seconds, 101.3 feet.

Any point on Earth can be specified using the pairing of latitude and longitude. The coordinates are always written with the latitude first, followed by a comma, and then the longitude.

By including minutes and seconds, the accuracy of such locations can be very precise. For example, the coordinates 32°45'33.5"N, 97°19'30.2"W point to an intersection in Fort Worth, Texas, and they are read as 32 degrees 45 minutes 33.5 seconds north and 97 degrees 19 minutes 30.2 seconds west.

In the exam, you may be presented with a chart but some of the latitude and longitude lines may not have a label. On U.S. charts, the adjacent to a labeled line is 30 minutes into the next or previous degree, and it's not labeled. There are tick marks representing 1 minute. So, in this case, you need to make deductive reasoning to find out what the label of the unknown lines. For example:

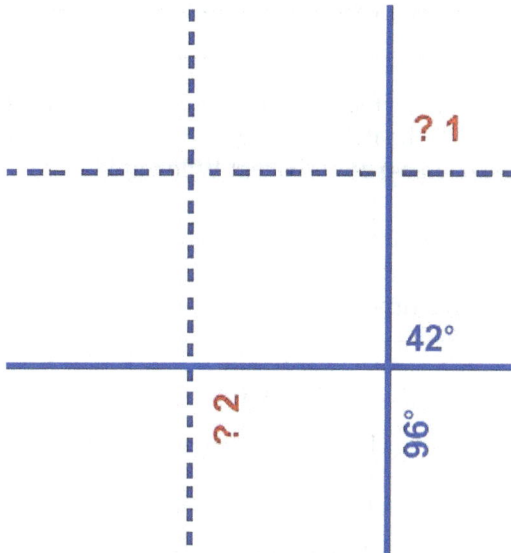

OK, so we have two variables to figure out, "?1" and "?2". First, figure out what kind of line is "?1". Just remember the horizontal lines would be parallel to the equator, so "?1" is a latitude line. Think of the latitude as a "la"dder rung. Since we are in the United States (above the equator) if we climb the ladder, the value will go up. Well, we have a rung with a value of 42, so "?1" would be between 42 and 43 degrees north, which is 30 minutes (half of a degree) so "?1" would be 42 degrees and 30 minutes north. Last, "?2" must be a longitude line. I don't want you to overthink the longitude line, for example, "Well, it's to the left, so it is a +1. But we are in the western hemisphere so that number is negative. When we add a positive one, wouldn't the negative get smaller?" This kind of overthinking will get you in trouble. Forget the number sign; just know that the number increases if you go to the left. So, to the left of 96 would be 96 degrees and 30 minutes west ("?2") since the next degree mark would be 97.

This is an actual FAA exam question from my last recurrency test:

---

1. Refer to FAA-CT-8080-2H, Figure 21.) What airport is located approximately 47 (degrees) 40 (minutes) N latitude and 101 (degrees) 26 (minutes) W longitude?

   A. Mercer County Regional Airport.
   B. Semshenko Airport.
   C. Garrison Airport.

OK, I don't want you to freak out since we haven't done chart familiarization yet. Also, realize that they placed the chart at a slight angle. Don't let this confuse you. What I do want you to do is look at the missing labels for latitude and longitude. Once we have that, logically, we should be able to answer this question. Let's bring up Figure 21 from the test supplement book.

So, on our "ladder rung," we step down one and figure out that the latitude below 48 degrees must be 47 degrees and 30 minutes north. With the longitude, we must go to the left, and the numbers get bigger as we recall. So, the longitude line to the left would be 101 degrees and 30 minutes west since the next mark would be 102 degrees west.

Now reading the question, we see that we are 40 minutes north on the latitude. Since we have a line on the 30-minute mark, we just estimate the point 10 minutes above it. We sketch dashes above the 30-minute mark. On the longitude, we are 26 minutes west, which is almost on the 30-minute mark. So now all we must do is intersect.

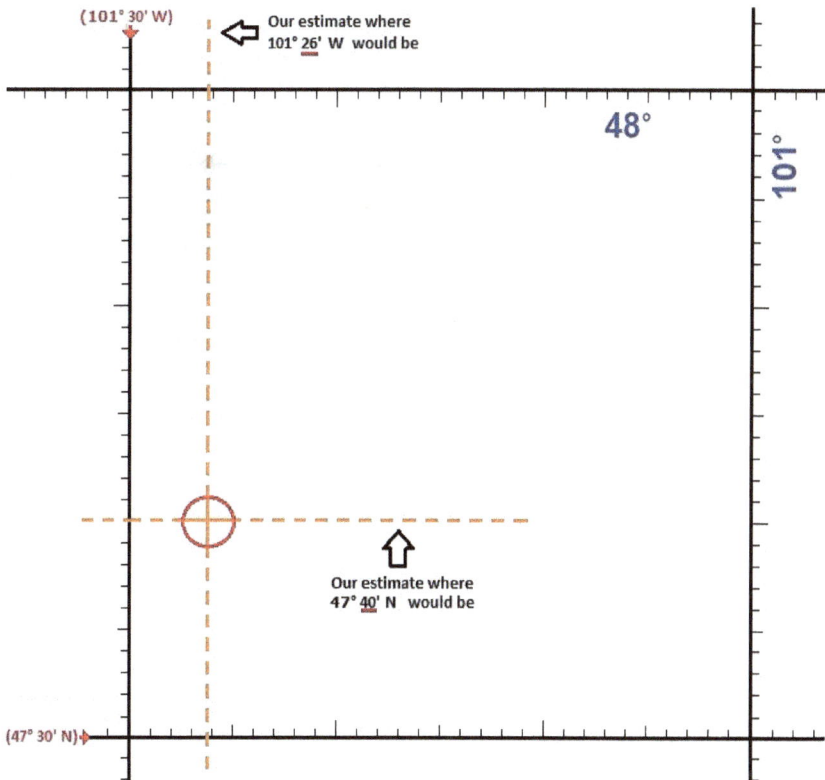

(101° 30' W)

Our estimate where
101° 26' W would be

48°

101°

Our estimate where
47° 40' N would be

(47° 30' N)

So, our target spot at the intersection would be approximately 47 (degrees) 40 (minutes) N latitude and 101 (degrees) 26 (minutes) W longitude. Looking at the chart, we noticed Garrison Airport to be in that vicinity.

As a recap, keep in mind that on coordinates and maps, minutes and seconds have nothing to do with time but rather a division of 1 degree:

1 degree = 60 minutes (approximate distance/degree 69 miles)

1 minute = 60 seconds (approximate distance/minute 1.15 miles)

1 second approximate distance is 101 feet.

Each degree of latitude and longitude "box" is divided into a quadrant of 30 minutes.

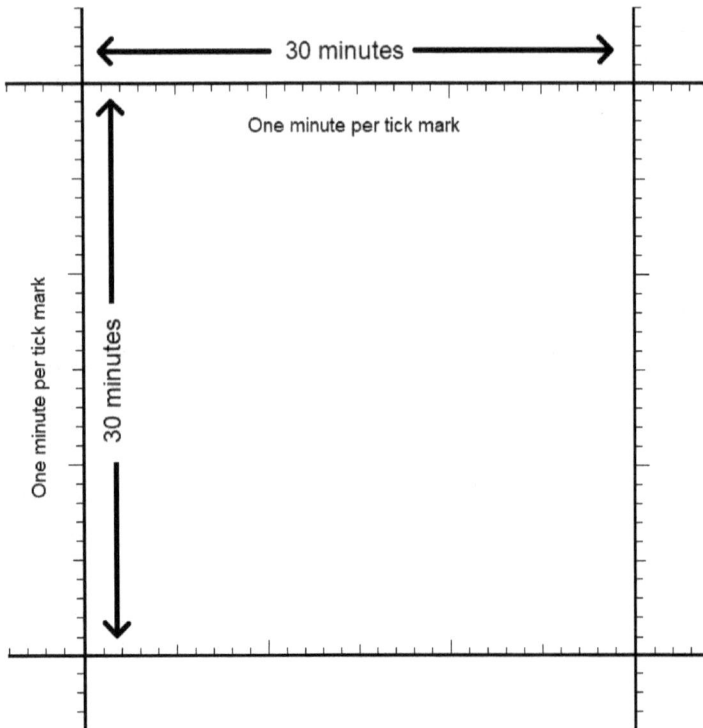

I have seen some questions in which the coordinates are given in terms of a fraction. For example:

What airport is at coordinates 47.9 N and 101.6 W?

I have seen the panic and the crazy math trying to figure out 0.9 and 0.6 in minutes. **Don't**. It doesn't matter how many minutes are in 0.9. You are given a percentage. Perfect. You know 0.6 of a dollar is 60 cents, right? So, all we must do is take the line from one degree to the next and simply estimate where 90% and 60% are located instead of counting tick marks. As long as you remember that the unlabeled line is 30 minutes (half a degree equals 50%). Sometimes giving a problem a pause, we may find a simpler solution.

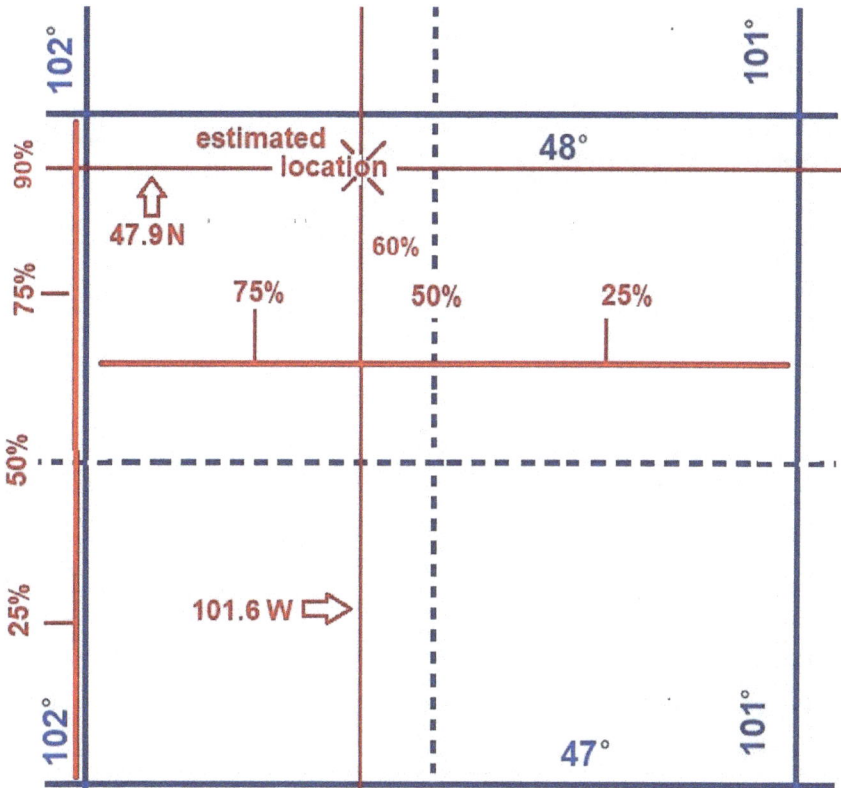

Altitude

The height of an object in the air is altitude. There are two types of altitudes: MSL (Mean Sea Level) or the height above sea level, and AGL (Above Ground Level). As far as drone pilots are concerned, our altitude is always AGL.

The altitude is measured by a device called an altimeter. This device relies on the barometric pressure which is the same as atmospheric pressure. It is a force exerted by the weight of air molecules in the Earth's atmosphere. It is usually measured in millibars (mb) or inches of mercury (inHg). At the standard sea level, the pressure is 29.92 inHg or 1013.2 millibars. As altitude changes, so does the atmospheric pressure. This is how an altimeter can read the changes in altitude based on the variations in atmospheric pressure. Aircraft pilots need to set their altimeters to the correct barometric setting for the area in order to get accurate readings. We will cover this more in the weather section. On a drone, the GPS supplies the height above ground level.

Speed

In aviation, speed is measured with knots. One knot is equivalent to 1.15078 mph. So, to convert knots to mph, you simply multiply with 1.15078. There are two types of speed: indicated and ground speed (the speed the aircraft is making over the ground below). An aircraft flying at 100 knots with a tailwind of 20 knots would have a ground speed of 120 knots. The opposite is true, flying 100 knots into a headwind would drop your ground speed.

Reading Charts

One of the challenging aspects of the test is references to aeronautical charts. It can be overwhelming and convoluted. Bring up the Figure 74 chart; you will understand what I mean. The best way to deal with charts is to understand the critical symbols we will need to get by on an exam. When taking the exam, put two pieces of paper in the FAA test supplement book. I have mentioned in the test description section to place a paper in the book's table of contents, the 7th and 8th page. The next one should be around the 13th page, listing the chart legends, symbols, and meanings.

The most used charts that would benefit us remote pilots are called the Sectional Charts. However, the exam test supplement book utilizes other charts, such as the VFR charts (for Visual Flight Rule). This aeronautical chart has a lot of information about the area you will be operating in. We will cover the most important symbols relating to remote pilot (and the exam).

Let's begin by bringing up the FAA test supplement according to this question (use the OneSite):

(Refer to FAA-CT-8080-2H, Figure 24, area 5.) At Sulphur Springs which frequency should be used as a Common Traffic Advisory Frequency (CTAF) to monitor airport traffic?

    A. 109.0 MHz.
    B. 118.35 MHz.
    C. 123.075 MHz.

Once the book is up, press Ctrl-F (find) and type "Figure 24". Press Enter until you see the chart. Once you have Figure 24, you have to rotate it until it is correct for reading orientation (on the test, you just rotate the book). Some browser allows a mouse to right-click, and "rotate clockwise" is in the popup menu. Once this is done, you must locate the airport in Area 5. The runway is usually the easiest to spot, next is the airport identifier. Since we are currently studying, we can review the symbols in this test supplement book. Use Ctrl-F (find) and type "legend 1." Include the period; otherwise, you'll find other unrelated "legends".

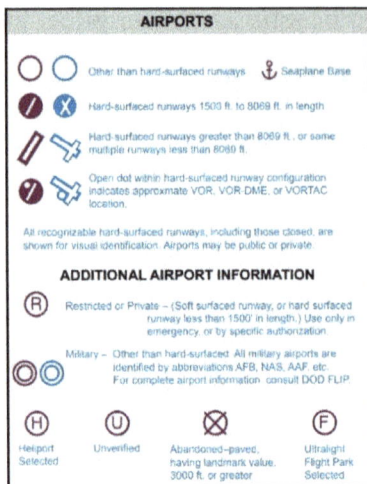

Let's look at the airport identifier. The first line should be the name and the location identifier (SLR).

The second line shows a radio frequency for the Automated Weather Observation System or AWOS. In large airports, the first line usually includes CT (the control tower frequency), the Automated Terminal Information System (ATIS), and others. The third line begins with a number in *italics*, the runway elevation in feet. So this airport is 489 feet above sea level (MSL). Followed by "L 50" which is the length of the longest runway in hundreds of feet. So this runway is 5,000 feet long. And trailing on line 3, we see a frequency in italics (123.075) followed by a symbol ⊙ ("C" in a dark circle). Consulting legend 1 symbols, we noticed that it "Indicates Common Traffic Advisory Frequencies (CTAF)" –bingo, we found our answer. Of course in the real exam, this would have taken only seconds because you know you will be looking for the C in the circle symbol for the CTAF frequency.

While we have Figure 24, let's look at other symbols we need to know. One of them is obstruction. On the chart obstructions can have many symbols:

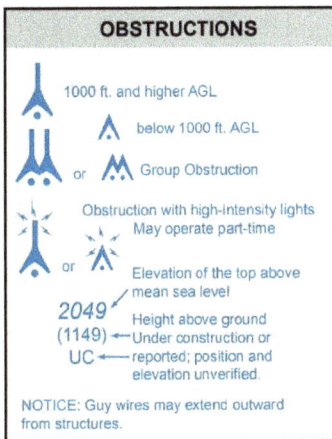

OBSTRUCTIONS

1000 ft. and higher AGL

below 1000 ft. AGL

or Group Obstruction

Obstruction with high-intensity lights
May operate part-time

or Elevation of the top above
mean sea level

2049 Height above ground
(1149) Under construction or
UC reported; position and
elevation unverified.

NOTICE: Guy wires may extend outward from structures.

Sometimes, several obstructions are grouped together. The height of the highest obstacle is displayed with it. The one listed on the top, usually in italics, is the height above sea level (MSL), and below it, usually in parentheses, is the height above ground level (AGL).

Let's look at a test question, keep your Figure 24:

(Refer to FAA-CT-8080-2H, Figure 24, Area 5.) A small UA is launched between the towns of Weaver and Saltillo east of Sulphur Springs airport. What is the height of the highest obstruction?

A. 914 feet MSL.
B. 344 feet AGL.
C. 355 feet AGL.

Looking at the chart, we see this between Weaver and Saltillo. It is a group of obstructions. So, the top number is MSL, and the bottom is 344 AGL. No other numbers listed on the choices fit the question.

The Maximum Elevation Figure (MEF) represents the highest elevation within a quadrant, including terrain and other vertical obstacles (towers, trees, etc.). A quadrant on Sectionals is the area bounded by ticked lines dividing each 30 minutes of latitude and longitude. MEF figures include additional height for possible error compensation and are rounded up to the nearest 100' value. The

last two digits of the number are not shown. The number to the left represents 1,700 feet MSL. See if you can spot MEFs on charts.

While we are on Figure 24, let's look at area 3 in the top left corner to address this question (note the parachute activity symbol and the

CTAF frequency used to monitor the activities in this parachute jumping area).

(Refer to FAA-CT-8080-2H, Figure 24, Area 3, and Legend 1.) For information about the parachute operations at Tri-County Airport, refer to

A. the notes on the border of the chart.
B. Chart Supplements U.S.
C. the Notices to Airmen (NOTAM) publication.

The Chart Supplements are the source whenever we need very detailed or in-depth information about airport operations. The notes on the border are for general info.

On this real test, you are on your own. See if you can figure it out.

(Refer to FAA-CT-8080-2H, Figure 78.) You have been contracted to inspect towers located approximately 4NM southwest of the Sioux Gateway (SUX) airport operating an unmanned aircraft. What is the maximum altitude above ground level (AGL) that you are authorized to operate over the top of the towers?

A. 400 feet AGL.
B. 402 feet AGL.
C. 802 feet AGL.

Check your answer with [107.51] on page 9.

When we get to the Airspace chapter, we will cover other symbols we need to know about.

Assignments:

1. **Review** the "Legend 1" figures in the Test Supplement book to familiarize yourself with the symbols listed.

2. **Review** the FAA Aeronautical Chart Users' Guide (use the OneSite).

This page intentionally left blank.

CHAPTER III

AIRSPACE

In the United States, the FAA regulates the airspace above you. There are four types of Airspace: controlled, uncontrolled, special use, and others. The complexity and the safety level required dictate the grouping. Here is the FAA Airspace classification:

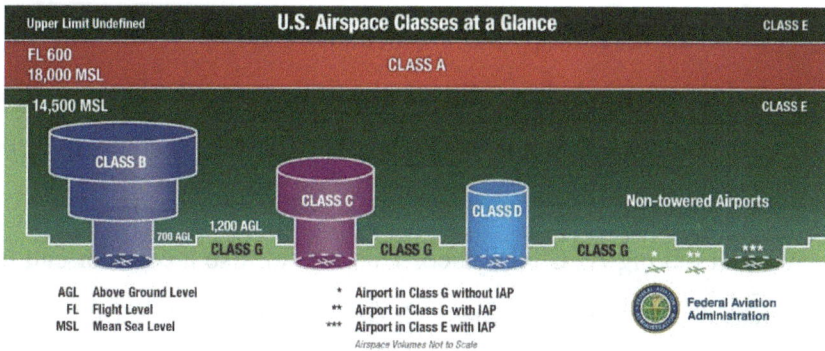

Controlled Airspace

There are four types of Controlled Airspace: Class B, C, D, and E. Each is unique with regulations and symbols on the aeronautical charts. One thing in common is that a remote pilot must obtain ATC authorization prior to flying within a controlled airspace.

Class B Airspace

Class B Airspace surrounds the nation's busiest airports. It consists of multiple layers, and it resembles an "upside-down wedding cake." This airspace is generally from the surface up to 10,000 feet MSL. It has all the latest technology, and its boundary is denoted with a solid blue line on the

Class B Airspace

charts. The floor and ceiling altitude of a class B Airspace is shown in solid blue figures with the last two zeros omitted. For example: $\frac{70}{SFC}$ This indicates from the surface (SFC) to 7,000 feet MSL. When the number is preceded by a plus(+) sign, this means "excluded."

$\frac{70}{+05}$ This means upward from above 500 to 7,000 feet (not including what's below). So, in the JFK example above, in that section, below 500 feet is *not* a Class B Airspace. Technically little planes and helicopters can sneak underneath 500 feet in that section without having ATC clearance. This is useful so a corridor can be "carved" underneath a Class B Airspace to allow traffic to fly without having to bother ATC. Tourists' helicopters flying around would be a good use of such a corridor. So, if you operate sUAS in such an area, it behooves you to have an individual serving as a spotter to let you know in the event of a small aircraft approaching your area.

Class C Airspace

Class C Airspace surrounds airports that are a step below class B but still have an operational control tower, with ATC able to accommodate Instrument Flight Rules (IFR) to include technology to facilitate such

flights. This airspace is generally 5 miles in radius, and it extends from the surface to 4,000 feet, with an outer circle extending from 1,200 to 4,000 feet above the airport elevation MSL. Its boundary is denoted with a **solid magenta** line on ▬▬▬▬ Class C Airspace the charts. The floor and ceiling altitude of a class C Airspace is shown in solid magenta figures with the last two zeros omitted. For

$\frac{70}{15}$ example: From 1,500 to 7,000 feet MSL. Reading the ceiling altitude may get a bit complex in many large airports with other nearby airports. Remember we discussed a Class B airspace that looks like an upside-down wedding cake? There are instances where a Class C airspace is sandwiched underneath a portion of the Class B airspace. This diagram is a side view of such an instance. In such a case, the ceiling altitude of Class C may not

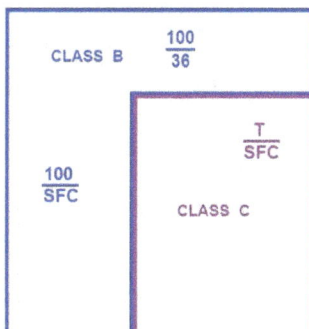

Side view of Class C sandwiched underneath a Class B layer.

have a number but instead the letter "T." A ceiling value of "T" indicates the ceiling is to, but not including, the floor of the overlying Class B airspace. (in this example, it is up to but not including 3600 –the base of Class B above the Class C airspace).

Class D Airspace

Class D Airspace surrounds airports that have an operational control tower. Each of the configurations of the Class D airspace area is individually tailored. Its boundary is denoted with a **blue** dashed line. This airspace is generally from the surface to 2,500 feet above the

— — — — Class D Airspace

airport elevation MSL. The ceiling is indicated in a blue dashed box with the last two zeroes dropped. When a minus sign is included in front, it is used to indicate "from surface to, but not including..."

Class E Airspace

Class E airspace is controlled airspace that is designated to serve a variety of terminal or en route purposes. Class E is rather tricky because its appearance differs depending on the usage. It is a controlled airspace that is neither A, B, C, or D. A very large airspace in

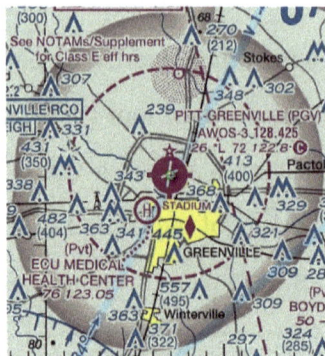

the U.S. is Class E. All airspace above 60,000 feet is Class E. For me personally, the color scheme was confusing. It can either be a gradient magenta or a blue gradient, or a segmented magenta! Ok, here is the deal, The inner circle of this picture is segmented magenta. Here Class E begins at the surface. — — — — — surface floor

floor 700 ft. above surface. Going out a bit we have a Gradient Magenta. With the Magenta Gradient: On the soft or fuzzy side of the line, Class E starts at 700' AGL. On the hard side of the line, it starts at 1200' AGL.

floor 1200 ft. or greater above surface that abuts Class G Blue Gradient: the fuzzy or soft side of the line, Class E starts at 1200' AGL. The heavy side abuts a Class G Airspace (next discussion).

2400 MSL Differentiates floors of Class E higher than 700 feet
4500 MSL above the surface.

By now some of you are wondering about **Class A Airspace**. In the United States, it extends from 18,000 feet up to 60,000 feet. While beyond the sUAS area of operation and not shown on visual charts, we need to know its existence. Incidentally, altitudes above 18,000 are referred to as "flight level" dropping the last two zeroes. For example, 22,000 is referred to as FL220 ("flight level two two zero").

Uncontrolled Airspace

In other areas outside the controlled airspace, it is the **Class G**. Class G Airspace in the United States extends up to 14,500 feet MSL. At and above this altitude is Class E Airspace. You don't need authorization to operate in Class G Airspace.

Special Use Airspace

Special Use Airspace or Special Area Operation (SAO) restricts entry of aircraft not part of the authorized activities. These areas are noted on the charts including area numbers. Several special use airspaces are as follows:

- Restricted areas.

- Prohibited areas.

- Warning areas.

- Military operation areas or MOAs.

- Alert areas.

- Controlled Firing Areas or CFAs.

Restricted Areas

These areas are hazardous to those not participating or authorized. While not completely prohibited, entry to such areas may be restricted unless you get permission from the using or controlling entities. These areas are noted with a boundary of blue hashed band and labeled as R-[number] such as R-658. Sometimes if the restricted area is not in use, the controlling agency may release the airspace to the FAA and ATC facility, allowing aircraft to operate in the airspace without issuing specific clearance.

Prohibited Areas

As the name implies, these areas are prohibited for aircraft not participating or authorized. Some of these areas are established for national security purposes. Unauthorized entry could be fatal. These areas are noted with a blue hashed band boundary and labeled as P—[number], such as P-40. Some Prohibited Areas also double as restricted areas. In such a case, they will display both the P and R numbers.

Warning Areas

This area may contain activities that may be hazardous to non-participating aircraft. It is usually located outward from the coast of the United States. These areas are noted with a boundary of blue hashed band and labeled as W-[number].

Military Operation Areas (MOAs)

This airspace was established to separate military activities from IFR aircraft. When in use, some IFR traffic may be authorized if ATC can provide proper IFR separation. Otherwise traffic will be rerouted. This area is denoted on the chart with a boundary of a magenta hash band and labeled as [name] MOA. The chart legend contains more specific information.

Alert Areas

These areas usually contain unusual activities, such as high-volume pilot training. Pilots should exercise extra caution in Alert Areas and be responsible for collision avoidance. These areas are also marked with a magenta hash band.

Control Firing Areas (CFAs)

This area is used as a controlled environment for some potentially hazardous activities. The difference between CFAs and other Special Use Airspace is that they usually suspend activities when a spotter sees an aircraft entering the area. It is unknown to this author if they will suspend activities for a small drone venturing into the area, so it is best to stay away. This area is not charted.

There are other airspace areas that may require special caution or restrictions. Some of these are:

Military Training Routes (MTRs) are used by military aircraft for training. They are established below 10,000 MSL. If the operation does not exceed 1,500 AG (e.g., VR1120, IR130), the routes are identified as IR (Instrument) or VR (Visual), followed by a four-digit number. Routes above 1,500 AGL are identified with a three-digit number (e.g., IR265, VR206). ◀— IR211

Parachute Jump Operation

The chart, which we discussed on page 46, denotes sites that are frequently used for parachute jump aircraft operation. It is published in the Chart Supplement U.S.

National Security Areas (NSAs)

These areas are defined with vertical and lateral dimensions established at locations where increased security is required. Flight into these areas may be temporarily or permanently prohibited by regulation under 14 CFR part 99 and is disseminated via NOTAM. If you use the FAA UAS Facility Map site, they are noted there, provided you include the layer "National Security." Here is a sample of such an area with UAS restriction.

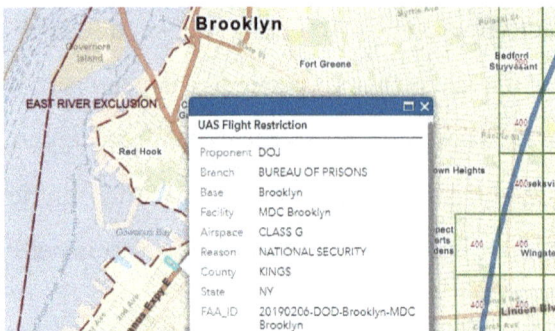

Temporary Flight Restrictions (TFRs) and NOTAMs.

These areas cannot be noted on charts due to their dynamic nature. It is published by the flight data center (FDC). It is extremely important that this be checked along with NOTAMs prior to *any flight*. TFRs are important due to the nature of what they protect, including the U.S. President and other public figures. Ever since September 11, 2001, the enforcement of TFRs has become much more stringent.

I cannot over-emphasize enough that TFRs are to be taken very seriously. There have been several incidents of aircraft incursions into TFRs that have resulted in an audition with the FAA for security investigations and certificate suspensions. NOTAM, on the other hand, deals with conditions that may affect the safety of flights. More permanent NOTAMs are published in print through a subscription from the Superintendent of Documents. There are different types of NOTAMS including DROTAMs (Drone NOTAMs) and BIRDTAMs (passage of a flock of birds in airspace).

Other special-use airspace information is beyond the scope of this book and UAS, such as Terminal Radar Service Area, Local Airport Advisory, and others related to VFR flights.

OK, so let's get started with some exam questions to practice. Bring up Figure 20 from the test supplement book. This is by far my favorite section of the map because it contains almost everything you can think of in terms of airspace and special use airspace. Along the way, we will learn some new symbols.

(Refer to FAA-CT-8080-2H, Figure 20, Area 1.) The Fentress NALF Airport (NFE) is in what type of airspace?

A. Class C.
B. Class E.
C. Class G.

That's a simple one. Clearly, it is dashed magenta, so it is Class E beginning at the surface.

(Refer to FAA-CT-8080-2H, Figure 20, area 2.) Why would the small flag at Lake Drummond of the sectional chart be important to a remote pilot?

A. This is a VFR check point for manned aircraft, and a higher volume of air traffic should be expected there.
B. This is a GPS check point that can be used by both manned and remote pilots for orientation.
C. This indicates that there will be a large obstruction depicted on the next printing of the chart.

Ah, we got a new symbol here...

LAKE DRUMMOND

A small flag indicates a checkpoint. There are several different types of checkpoints. One used for VFR, which is just a simple recognizable geographical feature, like a lake, railroad tracks, monument, etc. Pilots used this as a reference to make sure they were on the right track. On a non-controlled airport, they may broadcast the checkpoint over CTAF frequency (or other designated) so that other pilots can have situational awareness as to your location relative to their aircraft. For example, "Cessna November five two niner two Romeo heading one two zero over Lake Drummond..."

In the world of Instrument Flight Rule (IFR), where the outside view might be non-existent due to clouds, the checkpoint can be a navigational fix such as a waypoint. I'm sorry, but I'll digress to a non-drone aircraft... In such a case, the checkpoint would have other information like this one, which is a VOR checkpoint (VOR stands for VHF Omni-directional Radio range), a ground navigational aid transmitter used by pilots to navigate by tuning their radio to its frequency. In the above example (113.0), or channel 77 instead of scrolling through frequencies. The pilot can confirm that they are above the checkpoint by either their navigational equipment, or if they tuned their radio to 113.0 MHz, they would hear a Morse code (or voice on newer equipment) "dash dash dash, dash dot dot dot, and dash dot dash" (spelling OBK as the legend indicates). This confirms that they are indeed over OBK VOR. --OK, back to the exam question. So why would this visual checkpoint be of interest? Checkpoint would have, as choice A indicated; "... a higher volume of air traffic should be expected there."

Test question:

(Refer to FAA-CT-8080-2H, Figure 20, area 5.) How would a remote PIC "CHECK NOTAMS" as noted in the CAUTION box regarding the unmarked balloon?

A. By utilizing the B4UFLY mobile application.
B. By contacting the FAA district office.
C. By obtaining a briefing via an online source such as 1800WXBrief.com.

Looking at the chart, it looks like the caution box is south of Elizabeth City Regional Airport (ECG), abutting the VOR-DME ECG. So briefly I went to Get TFR-NOTAM from the OneSite, clicked on "NOTAM search" on the top navigation bar, and entered ECG. There were two NOTAMs that were listed, the top one being the VOR. I selected that, and sure enough, at the bottom of the list, there it is (emphasis added for clarity) "moored balloon."

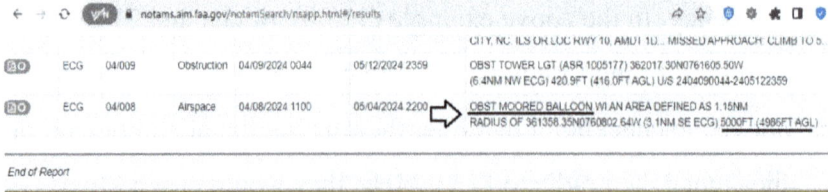

During the exam, it is important that you make a habit of reading the entire question and answers before processing. This question could have been answered without looking at Figure 20. The question was how do you look up NOTAMs? You would know it's an online source. So, if there is only one choice that is online, then it doesn't matter if it is faa.gov, 1800WXBrief.com, or whatever.

Assignments:

1. Using the OneSite, **Review** FAA Pilot's Handbook of Aeronautical Knowledge Chapter 15, Airspace.

2. Take the Knowledge Quiz for "Airspace" at the OneSite (see note below)

Airspace is one of the rather complex subjects. Please re-review this chapter for questions that you answered incorrectly during the quiz. if you score anything but perfect on the OneSite practice quiz, you need to go back and review.

This page intentionally left blank

CHAPTER IV

AVIATION WEATHER

Aviation weather observations and reports are developed to give an accurate depiction of current weather. Knowing the weather gives us the ability to understand what to expect when certain conditions arise. For the sUAS pilot, knowledge of current weather can be beneficial. For example, you don't want to fly if the wind gusts are too high because your drone will lose its stability in flight. The possibility of being thrown off course and hitting structures may result from windy conditions. Understanding how air flows around and above structures gives us the ability to anticipate. When flying approaching a tall structure, you can almost predict that you will encounter an updraft, then, when you clear the edge of the top edge of the building, a downdraft. In nature such as around mountains, these air movements are affected by temperature and barometric pressure. By understanding this we can expect a predictable flight. Flight during icing conditions would make the propeller less aerodynamic since an accumulation of frozen moisture on it causes drag and adds weight.

Because of this, a weather check should *always be* part of your preflight checklist.

Atmospheric Pressure

Air molecules around us exert pressure on anything it touches. The more air molecules, the more pressure. Look at a balloon; as you increase the number of air molecules in it, it exerts pressure on the inside wall, making it expand. To have a common denomination to perform calculations, there is a standardized base. The common denominator we use is the sea-level elevation. Weather observation stations, regardless of location and altitude, take air pressure readings and convert them to a value that would have been observed if that instrument was located at sea level. At sea level, the standard measurement for atmospheric pressure (ATM) at a standard 15 degrees Celsius is 1,013.25 millibars or 29.92 inches of mercury.

Dew Point and Air Temperature

The dew point is the temperature at which air becomes saturated with water vapor and starts to condense to water. When the air temperature is reduced to the dew point, the air is completely saturated and can no longer hold moisture. The moisture then begins to condense out of the air in the form of dew, fog, frost, clouds, rain, or snow.

Relative Humidity

The amount of water vapor present in the atmosphere at a given time is relative humidity. For example, if the current relative humidity is 70 percent, the air holds 70 percent of the total amount of moisture it can hold at that temperature and pressure.

Density Altitude

In addition to the regular altitude, which increases as you fly higher, there is also <u>Density Altitude</u>. Density altitude is affected by three things: Temperature, Altitude, and Humidity. So, depending on temperature and humidity, density altitude can change even if you are at the same altitude. Let's look at their relationship to density altitude: What reduces air density?

1. Higher temperature.
2. Higher humidity.
3. Higher altitude.

So even if the last variable (altitude) remains the same, the other two variables can make a difference. For example, on a hot and humid day, the air becomes "thinner" or less dense, and its density at the location is equivalent to a higher altitude in the standard atmosphere. Thus, the term "high-density altitude." High-density altitude **reduces aircraft performance** (the propeller has less to "bite" or grip. A jet engine has less air to ingest for combustion).

What effect does high density altitude have on the efficiency of a UA propeller?

A. Propeller efficiency is increased.
B. Propeller efficiency is decreased.
C. Density altitude does not affect propeller efficiency.

Just remember high density = bad performance.

Fog and Classifications

Fog is basically clouds that are near the ground level. When the temperature of air near the ground is cooled to the air's dew point. At this point, water vapor in the air condenses and becomes visible in the form of fog. Depending on how the fog came to be or formed, a classification with different names was established.

Steam fog, or sea smoke, is common over bodies of water. It is formed when cold, dry air moves over warm water. As the water evaporates, it rises and looks like smoke. It may cause low-level turbulence.

Radiation fog forms when the ground cools rapidly and the surrounding air temperature reaches its dew point. If radiation fog is less than 20 feet thick, it is known as ground fog. Radiation fog can usually be found in low-lying areas like mountain valleys. Ice fog is like radiation fog, but it occurs in extremely cold weather when the temperature is below freezing and water vapor forms directly into ice. It is usually only found in the Arctic regions.

Advection fog is most likely to form when a layer of moist, warm air moves over a cooler surface. Unlike radiation fog, wind is a required element to build the density of the fog. It is commonly found in coastal regions when the ocean's warm and moist breeze blows inland.

Upslope fog usually occurs on sloping landmasses, such as mountain ranges, when moist, stable air is forced up the sloping land feature. Wind is a required component for the formation and continued existence of this type of fog. Upslope fog can persist for days and extend to greater altitudes than radiation fog.

Atmospheric stability depends on its ability to resist vertical motion. In stable air, vertical disturbances are almost nonexistent. Unstable air comes when air is warm and moist and will bring on thunderstorms and turbulent air. These are the characteristics of stable and unstable air masses. (* It's a little counterintuitive.)

Stable Air	Unstable Air
Stratiform clouds and fog	Cumuliform
Continuous precipitation	Showery precipitation
Smooth air	Turbulence
Fair to poor visibility*	Good visibility*

Ceiling

In aviation, a ceiling is the lowest layer of clouds reported as broken or overcast or the vertical visibility into an obscuration like fog or haze. Clouds are reported as broken when five-eighths to seven-eighths of the sky are obscured. When the entire sky is obscured with clouds, it is called overcast. Current ceiling information is reported by automated weather reports such as METAR (discussed later).

Visibility

Visibility is the furthest horizontal distance at which objects can be seen with the naked eye. Current visibility is also reported in aviation weather reports. As for us, the visibility requirement for a flight of sUAS is three statute miles. Don't let test questions mentioning different airports and locations confuse you. It's always 3 miles.

Precipitation

Precipitation refers to any form of water, liquid or solid (snow or ice particles), that forms in the atmosphere and falls to the ground. It can reduce visibility, create icing situations, and affect anything that is flying, including drones. A buildup of ice on the propeller will cause it to be unable to maintain flight.

Thunderstorms

Thunderstorms go through three stages. The first is the cumulus stage, where rising begins after the air becomes unstable. As the clouds continue to increase in height with strong updrafts, within roughly 15 minutes, the thunderstorm reaches its mature stage, the most violent time period of its life cycle. At this stage, the cloud can no longer support the weight of the moisture. In this stage, drops of moisture, whether solid or liquid, begin to drop as rain or ice; this creates a downward motion of the air. Violent turbulence exists within and in the vicinity of the cloud. Under the clouds, the down-rushing air causes surface winds to develop and decreases the temperature. Once the vertical slows down, the top of the cloud widens in area. At this point, the storm enters the dissipating stage. This is when the downdrafts disperse and replace the updraft.

While this section of the book will by no means make you a meteorological expert, it should hopefully provide sufficient background information to recall during the exam.

Automated Weather Reports

Aviation weather reports are the results of a compilation of multitudes of networks that provide continuous weather information. The two automated reporting systems we need to know about are METAR (METeorological Aerodrome Report) and TAF (Terminal Aerodrome Forecast). While both deal with weather, METAR is the <u>current</u> observation and condition, while TAF is, as the name suggests, a forecast or <u>prediction</u> as to what the weather will be in the near future. Both systems provide cryptical messages regarding the weather. Fortunately for us, both systems use the same terminology so once you know one, the other is easy to understand. What we must learn is the decoding and terminology used. Before we continue, we need to pick up a couple of new things: Time and Temperature. The time in a METAR is given in 24-hour format UTC (Universal Time Coordinated), known as ZULU time. While not the same, Greenwich Mean Time (GMT) at the Prime Meridian, is the closest we can get to converting Zulu to local time:

UTC – 4 = Eastern Daylight Time (EDT) (summer)
UTC – 5 = Eastern Standard Time (EST)
UTC – 5 = Central Daylight Time (CDT)
UTC – 6 = Central Standard Time (CST)
UTC – 6 = Mountain Daylight Time (MDT)
UTC – 7 = Mountain Standard Time (MST)
UTC – 7 = Pacific Daylight Time (PDT)
UTC – 8 = Pacific Standard Time (PST)
UTC – 8 = Alaskan Daylight Time (ALDT)
UTC – 9 = Alaskan Standard Time (ALST)
UTC -10 = Hawaiian Standard Time (HST)

To convert 24-hour format, add AM if less than 12. Subtract 12 if greater than 12 and add PM. For example, convert 2100 Z to Eastern Standard Time: 21 – 5 = 16 EST. 16 – 12 = 4 PM. Another example is converting 1000 Z to Pacific Standard Time: 10 – 8 = 2 PST, 2 AM.

All temperatures in METAR and TAF are given in centigrade (Celsius). To convert Celsius to Fahrenheit, we use this formula: °F = (°C × 9/5) + 32. For example, 14° C would be (14° x 9 / 5) + 32 = 57.2° F.

A METAR report consists of elements separated by a space. When a space is encountered, a new element will be described.

How to Decode METAR Report

Here is a test question:

> (Refer to FAA-CT-8080-2H, Figure 12.) The wind direction and velocity at KJFK is from
>
> A. 180° true at 4 knots.
> B. 180° magnetic at 4 knots.
> C. 040° true at 18 knots.

Let's bring up Figure 12 from the test supplement book.

```
METAR KINK 121845Z 11012G18KT 15SM SKC 25/17 A3000

METAR KBOI 121854Z 13004KT 30SM SCT150 17/6 A3015

METAR KLAX 121852Z 25004KT 6SM BR SCT007 SCT250 16/15 A2991

SPECI KMDW 121856Z 32005KT 1 1/2SM RA OVC007 17/16 A2980 RMK RAB35

SPECI KJFK 121853Z 18004KT 1/2SM FG R04/2200 OVC005 20/18 A3006
```

Uh oh… we see METAR, but what on earth is SPECI?

While a METAR report is generated every hour or half an hour at most stations, a SPECI is a special observation message highlighting any changes or information added since the last issued METAR. Obviously, the report issue time becomes an important aspect for us in determining which one is the most current. As has been mentioned, the good thing is all these reports (METAR, SPECI, and TAF) are written and coded using the same terminology. The abbreviations are listed at the end of the chapter. Let's grab the third line and learn how to decode it before we look at the SPECI report.

METAR KLAX 121852Z 25004KT 6SM BR SCT007 SCT250 16/15 A2991

① ② ③ ⑤ ⑥⑦ ⑧ ⑨⑩

Seq	Field	Value	Interpretation
1	Type of report	METAR	Regularly scheduled METAR
2	Station ID	KLAX	LAX Airport Los Angeles, CA
3	Date & Time	121852Z	12th day, 18:52 UTC
4	Modifier	- Not included -	(Source of data e.g. AUTO)
5	Wind	25004KT	250° at 4 Knots
6	Visibility	6SM	6 Statute Miles
7	Weather	BR	Moderate Mist (-light, + heavy)
8	Sky condition	SCT007 SCT250	Scattered clouds at 700 & 2500 Ft
9	Temperature	16/15	Temp 16° C, dew point 15° C
10	Altimeter Set	A2991	29.91 inches of mercury

There is another element, RMK (Remarks), which was not included in this METAR. I don't want you to get hung up with memorizing the sequence, they may change when things are added to the report. What you need to do is focus on understanding the content of the elements of the report. So, without looking at the sequence, if you see an entry ending with KT then you know it is wind speed, ending with M, visibility in miles, ending with Z, Zulu time, etc. Once you can decipher this, the mystery disappears. For example, without thinking of the sequence, look at the KJFK SPECI report. Can you find the wind speed? Yes--18004KT (180° at 4 Knots). But wait, there are two answers, one is true, and one is magnetic. Which one is it? Not sure if you remember this from the crash course on page 33, but I'll repeat it:

If you can <u>read</u> it, and it is not about a compass, it's **true**. If you can <u>hear</u> it, it's **magnetic**.

You read it. The answer is A. 180° true at 4 knots.

The SPECI report has other elements not listed on the METAR above; for example, at the end of visibility and weather elements, it has "R04/2200", which is decoded to "Runway 04, the visual range is 2200 meters (yes M). We are fortunate with the current technology that in real life, no one decodes these anymore (try METAR/TAF decoder on the OneSite), but for study and test purposes, we need to be able to decipher these reports. The last one we will cover is TAF.

Terminal Aerodrome Forecasts (TAF)

Most larger airports have the TAF service to provide reports within a 5-mile radius of the airport. Each TAF is valid for a period of usually 24 hours. They are updated four times a day at 0000Z, 0600Z, 1200Z, and 1800Z. While the TAF report utilizes the same descriptors as METAR, it looks a lot more convoluted. We will try to break it down to make it simpler. Let's look at a test question:

> (Refer to FAA-CT-8080-2H, Figure 15.) What are the hours on day 12 where the probability for rain or snow is 40% at Memphis airport?
>
> A. Between 0500 and 0800Z.
> B. Between 1220 and 1222Z.
> C. Between 0200 and 0500Z.

Let's bring up KMEM TAF so we can dissect it.

```
KMEM 121720Z 1218/1324 20012KT 5SM HZ BKN030 PROB40 1220/1222 1SM TSRA OVC008CB
    FM122200 33015G20KT P6SM BKN015 OVC025 PROB40 1220/1222 3SM SHRA
    FM120200 35012KT OVC008 PROB40 1202/1205 2SM-RASN BECMG 1306/1308 02008KT BKN012
    BECMG 1310/1312 00000KT 3SM BR SKC TEMPO 1212/1214 1/2SM FG
    FM131600 VRB06KT P6SM SKC=
```

Let's decode the first block. You are familiar with these terms from METAR but there are some additional descriptors.

KMEM	Station ID - Memphis airport identifier
121720Z	Date and Time, day 12, 17:20 UTC
1218/1324	*NEW* Valid Time Period. Starts on day 12, 18:00 UTC and ends on day 13 at 24:00 UTC.
20012KT	Wind Direction 200°, speed: 12 KT.
5SM	Visibility 5 Statute Miles
HZ	Weather Haze
BKN030	Sky condition: Cloud Layer Broken (BKN) 3,000ft AGL

So, we have a new descriptor, but the rest looks like METAR. Here is a new descriptor that is not in METAR namely forecast or probability. It looks like this:

PROB40	*NEW* Probability Forecast 40% chance of weather event
1220/1222	Day 12 between 20:00 and day 12 @22:00 UTC

Now we understand this, all we must do is continue and see what the events are.

1SM	Visibility of 1 statute mile
TSRA	Thunderstorm(TS) Rain(RA)
OVC008CB	Sky outlook Overcast (OVC) 800ft AGL Cumulonimbus

There is another new group called the "From group". What this is saying is that from (for after) such and such time expect the following condition. It's coded like this:

FM122200	*NEW* Weather forecast for *after* Day:12, @22:00 UTC
33015G20KT	Wind Direction: 330°, Speed: 15KT, Gusts: 20KT

Returning to the question, we noticed that we must find a certain weather condition, rain, or snow. Looking at our abbreviation table, RA = Rain and SN = Snow --so anything with RASN.

KMEM 121720Z 1218/1324 20012KT 5SM HZ BKN030 PROB40 1220/1222
    FM122200 33015G20KT P6SM BKN015 OVC025 PROB40 1220/12:
    FM120200 35012KT OVC008 PROB40 1202/1205 2SM-RASN BEC
    BECMG 1310/1312 00000KT 3SM BR SKC TEMPO 1212/1214 1/2S
    FM131600 VRB06KT P6SM SKC=

We found it on the third line in the group of 40% probability (PROB40), so that matches. Now we find the times: It's 1202/1205, so day 12 from 0200 UTC to 0500 UTC. So, C is the answer.

METAR/TAF LIST OF ABBREVIATIONS /ACRONYMS

FMH-1 Federal Meteorological Handbook No.1, Surface Weather Observations & Reports.

ACC altocumulus castellanus	DOC Department of Commerce
ACFT MSHP aircraft mishap	DOD Department of Defense
ACSL altocumulus standing lenticular cloud	DOT Department of Transportation
AO1 automated station without precipitation discriminator	DR low drifting
	DS duststorm
	DSIPTG dissipating
AO2 automated station with precipitation discriminator	DSNT distant
	DU widespread dust
	DVR dispatch visual range
ALP airport location point	DZ drizzle
APCH approach	E east, ended, estimated ceiling
APRNT apparent	FAA Federal Aviation Administration
APRX approximately	
ATCT air traffic control tower	FC funnel cloud
AUTO fully automated report	FEW few clouds
B began	FG fog
BC patches	FIBI filed but impracticable to transmit
BKN broken	
BL blowing	FIRST first observation after a break in coverage at manual station
BR mist	
C center (runway)	
CA cloud-air lightning	FRQ frequent
CB cumulonimbus cloud	FROPA frontal passage
CBMAM cumulonimbus mammatus cloud	FT feet
	FU smoke
CC cloud-cloud lightning	FZ freezing
CCSL cirrocumulus standing lenticular cloud	FZRANO freezing rain sensor not available
cd candela	G gust
CG cloud-ground lightning	GR hail
CHI cloud-height indicator	GS small hail and/or snow
CHINO sky condition at secondary location n/a	HLSTO HLSTO hailstone
	HZ haze
CIG ceiling	IC ice crystals, in-cloud Lightning
CLR clear	
CONS continuous pellets	ICAO International Civil Aviation Organization
COR correction to a previously disseminated observation	

INCRG increasing	PY spray
INTMT intermittent	R right (runway) also runway
KT KNOTS	RA rain
L left (with reference to runway)	RMK remarks follow
LAST last observation before a break in coverage at a manual station	RTD Routine Delayed (late) observation
LST Local Standard Time	RV reportable value
LTG lightning	RVR Runway Visual Range
LWR lower	RVRNO RVR system values n/a
M minus, less than	RY runway
max maximum	S snow, south
METAR routine weather report provided at fixed intervals	SA sand
	SCSL stratocumulus standing lenticular cloud
MI shallow min minimum	SCT scattered
MOV moved/moving/movement	SE southeast
MT mountains	SFC surface
N north	SG snow grains
N/A not applicable	SH shower(s)
NCDC National Climatic Data Center	SKC sky clear
NE northeast	SLP sea-level pressure
NOS National Ocean Survey	SLPNO sea-level pressure n/a
NOSPECI no SPECI reports are taken at the station	SM statute miles
	SN snow
NOTAM Notice to Airmen	SNINCR snow increasing rapidly
NW northwest	SP snow pellets
NWS National Weather Service	SPECI an unscheduled report taken when certain criteria have been met
OCNL occasional	
OFCM Office of the Federal Coordinator for Meteorology	SQ squalls
	SS sandstorm
OHD overhead	STN station
OVC overcast	SW snow shower, southwest
OVR over	T precision temperature
P indicates greater than the highest reportable value	TCU towering cumulus
	TEMPO Temporary fluctuation within 30 to 60 minutes
PCPN precipitation	
PL ice pellets	TS thunderstorm
PK WND peak wind	TSNO thunderstorm information n/a
PNO precipitation amount n/a	
PO dust/sand whirls (dust devils)	TWR tower
PRES pressure	UNKN unknown
PR partial	UP unknown precipitation
PRESFR pressure falling rapidly	UTC Coordinated Universal Time
PRESRR pressure rising rapidly	V variable
PWINO precipitation identifier sensor n/a	

VA volcanic ash	WMO World Meteorological
VC in the vicinity	Organization
VIS visibility	WND wind
VISNO visibility at secondary	WSHFT wind shift
location n/a	Z zulu, i.e., Coordinated
VR visual range	Universal Time (UTC)
VRB variable	
VV vertical visibility	$ "$" is an indication that a sensor
W west	requires maintenance. Accuracy
WG/SO Working Group for	of report not guaranteed.
Surface Observations	= end TAF report

Assignments:

1. Using the OneSite, **Review** FAA Pilot's Handbook of Aeronautical Knowledge chapter 12 (Weather Theory).

2. On your own, use the OneSite and get METAR and TAF for the airport near you. Look at the decoding to become familiar with the terminology of the elements.

3. Take the Knowledge Quiz for "Weather " at the OneSite.

Weather is another rather complex subject. Please re-review this chapter for questions that you answered incorrectly during the quiz. If you scored anything but perfect on the OneSite practice quiz, you need to go back and review.

This page intentionally left blank.

CHAPTER V

OPERATIONS AND PERFORMANCE

In any Part 107 operation, the PIC is responsible for all aspects of flight, from preflight checks to shut down. There is not one portion of the flight operation for which the PIC is not responsible *(remember this for the test).*

Preflight Checklist

Before each flight, you must ensure that your aircraft is sound and capable of flying safely. As part of the preflight, the PIC should ensure that the sUAS is structurally sound and that nothing is cracked or broken. The propellers are in good shape and have no nicks and cuts, which might affect the performance. The battery is fully charged and in good condition. Pay attention to the battery casing and examine it to ensure there are no bulges or cracks. Lithium batteries can cause a serious fire hazard when damaged. You should not use a damaged battery for any flight. Note: When transporting a Lithium battery on board a commercial airliner, ensure you have it in your carry-on luggage so you can keep an eye on it instead of putting it out of sight in your checked-in luggage.

The next task on the checklist should be to ensure that your sUAS is properly balanced if it carries a load. Load balancing should be done to ensure that the center of gravity has not been adversely affected by loading. The center of gravity (CG) is not a fixed location on your sUAS but rather a location where there is an equal weight distribution depending on how the load is distributed.

If carrying a load, ensure that the sUAS does not exceed the maximum take-off weight specified by the manufacturer. Remember that even if you are within the manufacturer's envelope of load and weight, certain conditions, such as temperature, humidity, and density altitude, affect your sUAS performance.

If maintenance is to be performed, ensure that it is according to the manufacturer's recommended maintenance schedule. If one doesn't exist, you, as the PIC, must establish one.

[107.49]

You must ensure prior to any flight that the flight can be conducted safely including consideration of airspace, weather, location, ground hazard, battery level, informing others, good remote link, and other risks.

Load Factor

During flight, any maneuver, except for straight and level flight, will impose a load factor on the structure of the aircraft. In aerodynamics, a load factor is the ratio of the lift force to the weight of the aircraft. A load factor is measured in acceleration of gravity, or simply Gs, the force exerted by gravity. A load factor of 2 Gs, for example, means that the aircraft's structure is subjected to forces twice its weight. In a bank or turn, the aircraft will experience an increase in load factor. The load factor of a fixed wing aircraft is very different than an sUAS. In a fixed-wing aircraft, the wing acts as the airfoil to generate lift. It is very different from sUAS or helicopters, where the rotor blades generate lift from the pressure difference between the upper and lower blades. For the purpose of the exam, we need to understand how to calculate load factors on a fixed-wing aircraft. Let's look at this test question.

> (Refer to FAA-CT-8080-2H, Figure 2.) If an unmanned fixed-wing UAS weighs 33 pounds, what approximate weight would the aircraft's structure be required to support during a 30° banked turn while maintaining altitude?
>
> A. 34 pounds.
> B. 47 pounds.
> C. 38 pounds.

All we must do is multiply the load factor from the chart by the bank angle. Here is the load factor chart from the test supplement:

Here, we find the turn angle in the left column and multiply it by the adjacent load factor. So, 30° has a 1.154 load factor. So, 33 lbs. x 1.154 = 38.082 lbs. or answer C.

Angle of bank φ	Load factor n
0°	1.0
10°	1.015
30°	1.154
45°	1.414
60°	2.000
70°	2.923
80°	5.747
85°	11.473
90°	∞

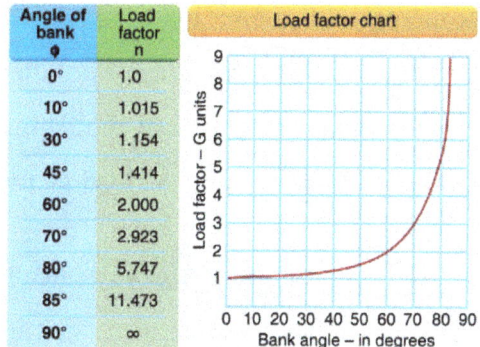
Load factor chart

An increased load factor also increases the risk of stalling. What is a stall? A stall is when an aerodynamic airfoil is no longer able to produce lift. Hence, the body no longer can stay in the air. An aerodynamic foil generates lift while it is flying through the air. An aircraft wing is an example of such an airfoil. A wing has an imaginary straight line that intersects the wing from the frontmost point to the trailing edge. This imaginary line is called the cord line.

The "angle of attack" is the angle of incidence between the incoming airflow and the wing's cord line. So, this is the angle of the wing as it "attacks" the incoming airflow. The lift can be increased by increasing the angle of attack. But this has a limit. If the angle of attack is too great, the airflow over the wing separates from the surface. Once this critical angle is exceeded, bad things happen. Namely, the smooth airflow over the wing becomes disrupted, and the wing stops producing lift. The name for this condition is a <u>stall</u>.

A study of this effect has revealed that a wing stall speed increases in proportion to the square root of the load factor. This means that an aircraft can be stalled by inducing a load factor. A pilot should be aware of the danger of inadvertently stalling the aircraft by increasing the load factor, as in a steep turn.

The pilot should always be aware of the consequences of overloading. An overloaded aircraft may not be able to leave the ground. If not properly loaded, performance suffers. Excessive weight reduces flight performance, including reduced battery life.

The PIC must understand the effect of weight on the performance of the particular aircraft being flown. Excessive weight reduces the safety margins available to the pilot.

Operations Above Human Beings

14 CFR 107 was amended in 2021 to include activities previously not permitted for sUAS; namely flight over human beings and night flight. We have covered the night flight in Chapter I.

[107.39]

No person may operate a small unmanned aircraft over human beings unless the person is directly participating in the operation or located under a covered structure unless it meets the operational requirement of Subpart D (explained below).

Subpart D deals primarily with flights over human beings or vehicles which was not present in the original 2016 CFR. We just aren't allowed then to fly over people. This subpart breaks down the UAS into four categories. Regardless of the weight of your drone, if it does not have blade/propeller protection, it will not qualify to be operated under these categories. The reason is that each category states that the drone "Does not contain any exposed rotating parts that could lacerate human skin upon impact with a human being."

If your UAS fits any of the categories and you intend to use it over people, the drone must be labeled appropriately with the Category it falls under.

Let's review based on [107.100] category operations over human beings.

[107.100]

Cat	Weight/Impact	Regulation
1	Weighs 0.55 pounds or less.	[107.110] No exposed blades.
2	Weighs over 0.55. Will not cause injury to a human being. The impact force of 11 foot-pounds of kinetic energy.	[107.115] - [107.120] No exposed blades. Listed in FAA acceptance declaration of compliance. No safety defects. Labeled Cat 2.
3	Weighs over 0.55. Will not cause injury to a human being. The impact force of 25 foot-pounds of kinetic energy.	[107.125] - [107.130] No exposed blades. Listed in FAA acceptance declaration of compliance. No safety defects. No open-air assemblies of human beings without RID. Labeled Cat 3.
4	Must use a small unmanned aircraft that is eligible for Category 4 Inspected with preventative maintenance.	[107.140] Have an airworthiness certificate issued under part 21 of this chapter. No open-air assemblies of human beings without RID.

[107.135]

Suppose a Category 2 or 3 label affixed to a small unmanned aircraft is damaged, destroyed, or missing. In that case, a remote pilot in command must label the aircraft in English so that the label is legible and prominent. The label will remain on the small unmanned aircraft for the duration of the operation over human beings. The label must correctly identify the category of operation over human beings that the small unmanned aircraft is qualified to conduct in accordance with this subpart.

Emergency Procedures

As mentioned before, the PIC is responsible for all phases of the flight, including emergency procedures and contingency plans, which must be reviewed together with your crew if one is available. For example, what will be done if the sUAS has lost communication with the remote control? Fly-aways because of GPS failure? In an emergency, you can deviate from the regulations in 14 CFR part 107 if your action is <u>necessary for safety reasons</u>. This deviation, however, must be reported to the FAA if requested.

Communications

If you operate within the vicinity of an airport, it is recommended that you monitor the CTAF or tower frequencies (arrival and departures at larger airports) to increase your situational awareness. Although you, as an sUAS pilot, are not expected to participate in or transmit over these frequencies, it is important that you be familiar with the aviation lingo used by controllers and pilots.

The International Civil Aviation Organization (ICAO) is a specialized agency that oversees global civil aviation operations and is responsible for providing uniform regulations, standards, and procedures. ICAO has adopted a phonetic alphabet that is used in radio communications. When communicating with ATC, pilots use this alphabet to identify letters and numbers:

ICAO Phonetic Alphabet and Numbers

A	Alpha (AL fah)	B	Bravo (BRAH VO)
C	Charlie (CHAR lee)	D	Delta (DELL tah)
E	Echo (ECK oh)	F	Foxtrot (FOKS trot)
G	Golf (GOLF)	H	Hotel (hoh TELL)
I	India (IN dee ah)	J	Juliet (JEW lee ET)
K	Kilo (KEY loh)	L	Lima (LEE mah)
M	Mike (MIKE)	N	November (no VEM ber)
O	Oscar (OSS car)	P	Papa (pah PAH)
Q	Quebec (keh BECK)	R	Romeo (ROW me oh)
S	Sierra (see AIR rah)	T	Tango (TANG go)
U	Uniform (YOU nee form)	V	Victor (VIK tah)
W	Whiskey (WISS key)	X	X-Ray (EKS RAY)
Y	Yankee (YANG key)	Z	Zulu (ZOO loo)

1	One (WUN)	2	Two (TOO)
3	Three (TREE)	4	Four (FOW-er)
5	Five (FIFE)	6	Six (SIX)
7	Seven (SEV-en)	8	Eight (AIT)
9	Nine (NIN-er)	0	Zero (ZEE Ro)

At an airport without a control tower, we must monitor the Common Traffic Advisory Frequency or CTAF. The CTAF is where the pilots announce their position and intentions. The CTAF might be a UNICOM, MULTICOM, FSS, or tower frequency identified on the chart. A MULTICOM frequency of 122.90 MHz will be used at a non-towered airport without an FSS or UNICOM.

Vision

For safety reasons, a remote pilot must always scan the area where they operate a small UA. This is especially important around an airport. To detect possible conflict or other traffic in the area, the PIC should use a scanning technique that systematically focuses on different segments of the sky for a brief interval and moves to the next segment systematically; for evening operations, avoid direct exposure to bright lights to maintain your night vision. The PIC must know that it takes approximately 30 minutes for a person's eye to adapt to darkness fully.

Physiological Factors Affecting Pilot's Performance

14 CFR [91.17] made it clear that no crew member may be involved with the flight operation if they had consumed alcoholic beverages within the preceding 8 hours, nor can they if their blood alcohol level is .04 percent or greater. We must realize, however, that there are many additional factors that may influence our ability to perform. Among others are standard over-the-counter medications such as antihistamines. Physical conditions such as fatigue after a long drive

can be just as deadly as being under the influence of drugs or alcohol. The bottom line is that the PIC must ensure that the operation can be conducted safely.

Aeronautical Decision Making (ADM)

ADM is decision-making in a unique environment, namely aviation. This discipline is very unforgiving for mistakes that might happen. Despite all the latest technology and safety devices, one variable remains the human factor. We make mistakes; there is no question about it. It is estimated that 80% of aviation accidents are related to human error.[2]

ADM is a systematic approach to risk management and mitigation that helps the stakeholders, such as crew and pilots, to be able to make sound decisions. ADM, in addition to Crew Resource Management, has positively impacted aviation safety in general. Crew resource management (CRM) training for flight crews focuses on effectively using all available resources: human resources, hardware, and information supporting ADM to facilitate crew cooperation and improve decision-making. The goal of all flight crews is a good ADM, and the use of CRM to aid in making good decisions. The risk management aspect, proactively identifying safety-related hazards and their mitigation, has become an important part of ADM. When followed and included in the pilot's decision-making process, it can reduce the inherent risk.

---

[2] Safety Science, Volume 140, August 2021, 105272

The importance of ADM skills includes our ability to recognize personal attitudes, which, when left unrecognized, may become a source of unsafe operation. Studies have identified five hazardous attitudes that can interfere with the ability to make sound decisions and exercise authority properly.[3]

Five hazardous Attitudes	Antidote
Anti Authority: "Don't tell me what to do!" This attitude is found in people who do not like anyone telling them what to do. In a sense, they are saying, "No one can tell me what to do." They may be resentful of having someone tell them what to do or may regard rules, regulations, and procedures as silly or unnecessary. However, it is always your prerogative to question authority if you feel it is in error.	**Follow the rules. They are usually right.**
Impulsivity: "Do it quickly." This is the attitude of people who frequently feel the need to do something, anything, immediately. They do not stop to think about what they are about to do, they do not select the best alternative, and they do the first thing that comes to mind.	**Not so fast. Think first.**
Invulnerability: "It won't happen to me." Many people falsely believe that accidents happen to others, but never to them. They know accidents can happen, and they know that anyone can be affected. However, they never really feel or believe that they will be personally involved. Pilots who think this way are more likely to take chances and increase risk.	**It could happen to me.**

---

[3] DOT/FAA/PM-86/41 report, May 1987.

Five hazardous Attitudes	Antidote
Macho: "I can do it" Pilots who are always trying to prove that they are better than anyone else think, "I can do it—I'll show them." Pilots with this type of attitude will try to prove themselves by taking risks in order to impress others. While this pattern is thought to be a male characteristic, women are equally susceptible.	**Taking chances is foolish.**
Resignation: "What is the use?" Pilots who think, "What's the use?" do not see themselves as being able to make a great deal of difference in what happens to them. When things go well, the pilot is apt to think that it is good luck. When things go badly, the pilot may feel that someone is out to get them or attribute it to bad luck. The pilot will leave the action to others, for better or worse. Sometimes, such pilots will even go along with unreasonable requests just to be a "nice guy."	**I'm not helpless. I can make a difference.**

During each flight, the PIC must continuously assess and mitigate risks. A useful tool is the *PAVE* checklist. By incorporating the *PAVE* checklist into preflight planning, the pilot divides the risks of flight into four categories: **P**ilot-in-command (PIC), **A**ircraft, en**V**ironment, and **E**xternal pressures (PAVE).

P = **P**ilot-in-command (PIC). The PIC is one of the risks and must evaluate the personal readiness for the flight.

A = **A**ircraft, is the aircraft appropriate, ready, and safe?

V = enVironment, airspace, weather, visibility, and temperature for the flight. Check TFR, NOTAMs.

E = External pressures, is the PIC pressured to make this flight? Does the PIC have something to "prove"?

Airport operations

An airport is a place used or intended for takeoff or landing, including the associated facilities, such as control towers, hangars, runways, etc. It also includes Heliports and seaplane-based terminals.

There are towered and non-towered airports, which are further subdivided into government/military airports, civil airports, and private airports.

As the name implies, a towered airport has a control tower manned by air traffic controllers. Non-towered airports, on the other hand, have no operational tower, and two-way radio communications are not required, although it is strongly recommended that pilots use CTAF to communicate their intentions.

When sUAS pilots operate near any airport, it is important that the PIC retrieves all available information for the airport. The most comprehensive airport informational resource is the "Chart Supplement U.S.," which includes a sectional chart, current NOTAMs, and TFR. Exercise caution for radio towers or antennas that may be supported by guy wires, which are difficult to see in good weather and almost invisible in low-light or bad weather conditions.

Best Practices

I am closing the operations and performance section with best practices to provide you with hints and tips that hopefully can be useful whether you want to fly for hire as an independent consultant, as part of a company, or even to open your own business.

LAANC Drone Airspace Approval
Check Airspace Before You Fly

Get a LAANC app. This app will give you more than the ability to file for authorization from ATC but also give you briefings for weather and area notices. There are many free applications that you can download from the app store. Double-check to ensure it is listed, recommended, or approved by the FAA. Most of these are free because the developers received subsidies from the government during the initial development of the app. I started with AirMap in 2016 but moved on to Flight Ready LAANC when AirMap went out of business.

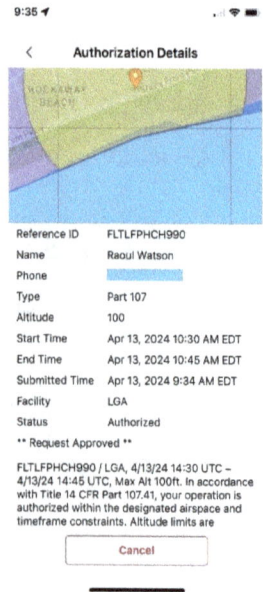

Get a logbook. To begin with, understand that you will become part of an industry where accountability is held high. Whenever something goes wrong, you will have people who will scrutinize everything you do. It is, therefore, important to always cover all bases to protect yourself.

Although the FAA does not require a logbook, I strongly recommend starting a logbook on day one of flying drones. The logbook should be simple and can be made with a simple binder with loose leaves and three-hole punch paper. I chose to do my logbook electronically by simply typing it on the computer and keeping a separate file for each flight. The benefit of this is that the files are time-stamped and not as easily faked as the written log. Next, you have to decide what to put in the logbook. Everything, that would be my answer. How else could you prove that you have done maintenance and that you have checked TFRs and NOTAMs? How would you prove that you have done your preflight? Write it down. At a minimum, your logbook should have the drone's FAA registration, ATC approval number if one is obtained, time, duration of flight, whether a visual observer is used, a spotter if used, the drone make and model, weather details, and your activities to include connectivity checks. Include a short drone video file name that you save somewhere to prove the drone view of your work area at the time of your activities.

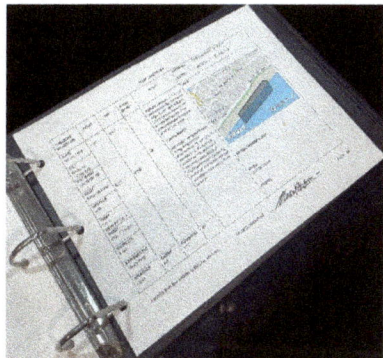

Get drone liability insurance. Your drone may crash and damage property or injure someone. Make sure you have insurance from a reliable drone or aviation-related insurance company. Have sufficient coverage because if you want to get hired by a company, for example, they may demand that you have at least $1 million in coverage. Do diligent research before you decide.

If you are planning to open your own company, consult a lawyer who knows the business (see below).

If you are flying drones commercially or have a company with Remote Pilots as employees, sooner or later, you will have to deal with either getting sued, needing assistance obtaining an FAA waiver, or worse, getting in trouble with the FAA. In such a case, you need more than a regular lawyer. You need an attorney who is very knowledgeable about the law regulating drone pilots and aviation. If I ever find myself in such a situation, one of the lawyers I would want on my side is Jonathan Rupprecht. Jonathan is a well-respected attorney in the industry. Because he is an aircraft pilot and FAA-certified Flight Instructor himself, his knowledge is far superior to that of regular attorneys. His website is https://jrupprechtlaw.com.

Assignments:

1. Using the OneSite, **Review** FAA Pilot's Handbook of Aeronautical Knowledge Chapter 2 (Aeronautical decision-making (ADM))

2. Take the Knowledge Quiz for "Operations" at the OneSite

This page intentionally left blank

CHAPTER VI

GETTING READY FOR THE TEST

This section will explain how to register, apply, and obtain your FAA Remote Pilot certificate. It also includes a section about the exam.

You will need to create an account on several sites to register and apply to take the test.

The first site is the Integrated Airman Certification and Rating Application (IACRA). Create an account and profile on this site and it will issue you a Federal Tracking Number (FTN). Please don't lose this number since you always need it on the FAA sites. You can use the OneSite for links.   https://iacra.faa.gov/IACRA/

New User select "Register"

The role you need to select during registration is "Applicant."

In the next section, you will ignore the certificate number (since you don't have one yet) and continue to fill in your personal information. Make sure you use your legal name **exactly** as it appears on official documents, such as your **passport or driver's license**. This is exactly how your name will appear on your test and certificate.

Please continue to complete all the required information for your account, create a user ID and password, and submit the registration.

If all goes well, you will receive an email confirming your registration. This email will also give you your assigned FTN (Federal Tracking Number). Don't lose this!

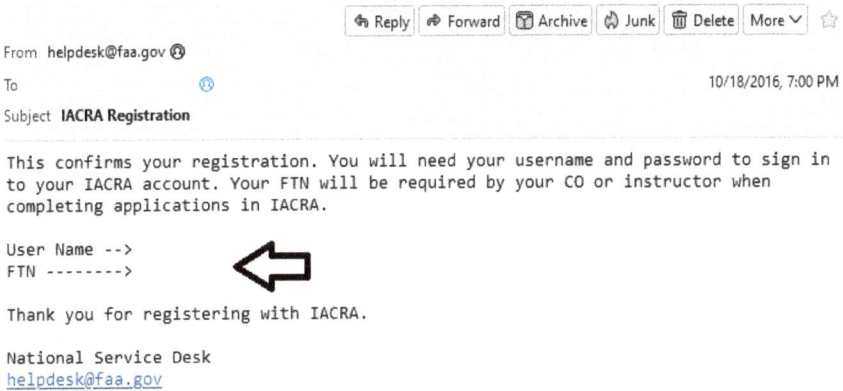

↩ Reply	↪ Forward	🗁 Archive	⟳ Junk	🗑 Delete	More ∨	☆

From helpdesk@faa.gov ⊗

To   ⊗   10/18/2016, 7:00 PM

Subject **IACRA Registration**

```
This confirms your registration. You will need your username and password to sign in
to your IACRA account. Your FTN will be required by your CO or instructor when
completing applications in IACRA.

User Name -->
FTN ------->              ⇐

Thank you for registering with IACRA.

National Service Desk
helpdesk@faa.gov
```

Once you have your FTN number, you must register and apply for the exam at one of the FAA Test Centers. This is accomplished by creating an account on the PSIexams site:

https://faa.psiexams.com/faa/login

You must create an account and fill out the eligibility form. You need your FTN number for this.

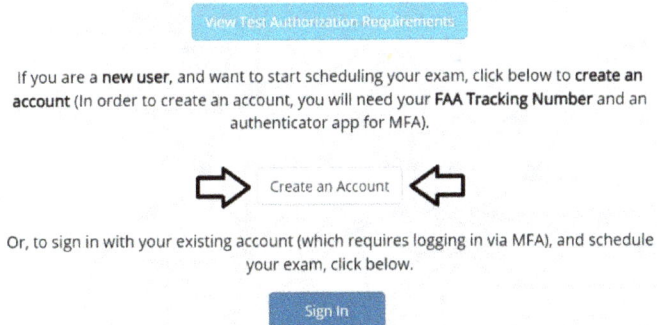

View Test Authorization Requirements

If you are a **new user**, and want to start scheduling your exam, click below to **create an account** (In order to create an account, you will need your **FAA Tracking Number** and an authenticator app for MFA).

⇒ Create an Account ⇐

Or, to sign in with your existing account (which requires logging in via MFA), and schedule your exam, click below.

Sign In

Verify Your Eligibility

An Integrated Airman Certification and Rating Application (IACRA), which includes your FAA Tracking Number (FTN), is required, in order to continue. The information entered must be an *exact match* with the information in IACRA.

The IACRA Website is the only source for creating or locating your FAA Tracking Number. Click here to go to the IACRA website.

FAA Tracking Number *(required)*

Enter FAA Tracking Number

First Name *(required)*

Enter First Name

Last Name *(required)*

Enter Last Name

Continue

Next, you would select the test "Unmanned Aircraft General (UAG) from the list and click "Schedule Exam". Select the test center which location is convenient for you and proceed to pay for the registration.

While you are creating accounts, there is one link that was listed on the OneSite, namely "FAA Safety Team (Courses and Webinars)." Create an account here since this is the site you must return in two years for your recurrency (you will be taking the course "ALC-677").

For now, you should take advantage of this site and look at the course catalog (it's all free), and you may be able to take courses related to Part 107 that can help you some more. *I recommend ALC-42, 43, 322, 449 and ALC-451.*

Federal Aviation Administration

Activities, Courses, Seminars & Webinars — Maintenance Hangar Pilots Resources

Activities | Courses | Seminars & Webinars | Topic Suggestions

Course Overview

Part 107 Small UAS Recurrent

Description: This course applies to any part 107 certificated remote pilot to update their 24 month recency of aeronautical knowledge. Completion of this course satisfies the 24 month recency of aeronautical knowledge requirement (currency) for a part 107 remote pilot with a sUAS rating. If you have any questions, please email the UAS Support Center at uashelp@faa.gov

Cost: No

Course Overview: Introduction
Part 107 sUAS Recurrent – Non-61 Pilots
Resources
Glossary
Review

The Day of The Test

On the day of the test, I can't really tell you what will work for you, but this is what I did back in 2016. If the info is useful, use it.

I had a good night's sleep and ate a very light breakfast. I was afraid I would have to go to the bathroom and waste my precious time. I promised myself that I would sit in the exam room and not get up until I finished. My test was scheduled for 10 am, but I got there early. I was in line with many other pilots; some took their exams for advancement. The lovely lady beside me was taking her CFI (Certified Flight Instructor). No one ever heard of part 107 and thought I was special. The proctor went to introduce himself in the room where we were told to assemble. I was given the test supplement book and a pencil with some blank paper. I told him I brought a calculator and a magnifying glass. He examined the calculator to ensure it was the permissible type and frowned upon my magnifying glass but had no objection. My phone and other belongings had to be secured in a locker.

I can tell you what... I was very happy to have brought a magnifying glass. The graphics on aeronautical charts have some fine print that is very difficult to read. So, if you wear glasses, bring them --but a magnifying glass will save the day.

While the proctor seemed disinterested while we were taking the exam, I noticed the location of a hidden camera. We were no doubt, monitored and watched.

During The Test

Ensure you read the questions and answers before attempting to find the correct answer. This has saved me numerous times by not having to flip pages of the test supplement book. For example, look at this question:

(Refer to FAA-CT-8080-2H, Figure 26.) What does the line of latitude at area 4 measure?

A. The degrees of latitude east and west of the Prime Meridian.
B. The degrees of latitude north and south of the equator.
C. The degrees of latitude east and west of the line that passes through Greenwich, England.

I was just about to flip pages to find Figure 26 and Area 4 when I realized that the question had nothing to do with whatever was on the chart's figure. It simply asks what latitude measures. We know that without even looking at the picture.

So again, read first and analyze.

If You Don't Know the Answer

Guess. You have one out of three chances of being right. Don't leave a question blank; you will lose a point.

Guess *but* be logical and choose an answer that makes sense to you *after* eliminating what doesn't make sense.

The answer is usually the PIC whenever the question says, "Who is responsible…".

Anytime they mention things for safety or to prevent injuries, go for that choice.

Finally, be efficient. You have two minutes to answer a question, so if you already spent a minute and still can't find the answer, there is nothing wrong with skipping the question. You can always come back later. The program does keep track of what is completed/answered and what is not.

Don't read too much into the questions. Take it at face value.

Be careful with numbers, especially 55. Since 55 is both the upper and lower limit of the sUAS class. I have used 0.55 in the text of this book, but on the exam, you may see .55. It is easy to misread .55 with 55.

At the end of the exam, you'll know your grade right away because the program scores it, and the proctor will give you a score printout. Note the exam ID number, which is significant for later.

If you pass, congratulations, but if you don't, it's not a big deal. Don't be discouraged! There is no limit to how many times you can retry. You do have to wait two weeks before you can reapply. Study some more and try again.

Shameless plug: If you pass on your first try, be kind enough to return and give this book a favorable review.

Submitting Test Results and Applying for The Certificate.

Once you pass, you must apply for your Remote Pilot certification. To do this, you must wait *at least* two full days to give a chance for the test center to submit your score to the FAA. Remember that when you leave the testing center, you should write down the exam ID number you took. This number will be needed for the application.

After at least 48 hours, log in to the IACRA site.

https://iacra.faa.gov/IACRA/

After you log in, accept the TOS and select "Start Application." On the application page, select "Pilot" and under the certificate category, "Small Unmanned Aircraft System".

Click on "Start Application"

On the next screen, please take your time to make sure what you enter is correct. You must use the same ID you use at the test site. If you use a driver's license, then that is the info you would enter. If you use your passport, then enter the passport information. The other critical entry is the "Knowledge test exam ID." Double-check to ensure these numbers are entered correctly. Don't rush.

Others to fill out are questions about whether you ever had a certificate denied and whether you have a physical challenge that prevents you from exercising the certificate.

That's pretty much it. Submit the application, and after 4 or 6 weeks, you should get the physical remote pilot card in the mail. In the meantime, you may receive an email stating that your temporary certificate is available. In such a case, you log back into IACRA and print out your temporary certificate.

I hope you will be as excited as I was when I received my certificate. It is indeed an accomplishment to be certified by the FAA.

Final Assignment:

Take the final quiz "PART 107 - All subjects" at the OneSite. It is timed as the FAA Part 107 test. Take this quiz several times with a day or two breaks since the database has more questions than the 60 presented randomly. Use this to gauge your readiness for the test. Give yourself a higher bar let's say 85% passing.

AFTERWORD

In the end, you can be proud of yourself. While not difficult, the material you covered was not for the faint-hearted. It was a lot of material and time-consuming to study. At this point, however, you have a new skill set that you can be very proud of.

The drone industry is rapidly growing. Currently, sUAS are used in commercial applications, such as photography, real estate, mapping, and inspection. They are also used in agriculture, the oil and gas industries, and law enforcement. Since its inception, it has only climbed in use and revenue. It is estimated to have a market value of $5.8 billion. The industry is projected to be worth over ten billion dollars by 2030.[4] Indeed, it is an exciting industry to be a part of.

The industry is very dynamic. Because of this, I can see that as rules and regulations change, there will be a new edition of this book. I am a good listener, so please review this book and let me know what you like and don't like. I can assure you that every single suggestion will be addressed. Thank you for choosing this book; good luck and best wishes. I'll see you in two years when you review the chapters before your recurrency test.

---

[4] Research & Markets, February 2024, 4623164

www.ingramcontent.com/pod-product-compliance
Lightning Source LLC
Chambersburg PA
CBHW051431090426
42737CB00014B/2913